A First Course in "In Silico Medicine"

Volume 2

Series Editor
Masao Tanaka
Professor of Osaka University

1-3 Machikaneyama, Toyonaka
Osaka 560-8531, Japan
tanaka@me.es.osaka-u.ac.jp

T0214971

Shinji Doi · Junko Inoue · Zhenxing Pan
Kunichika Tsumoto

Computational
Electrophysiology

Dynamical Systems and Bifurcations

 Springer

Shinji Doi
Professor
Graduate School of Engineering
Kyoto University
Kyoto-Daigaku Katsura, Nishikyo-ku
Kyoto 615-8510, Japan
doi@kuee.kyoto-u.ac.jp

Junko Inoue
Associate Professor
Faculty of Human Science
Kyoto Koka Women's University
38 Kadono-cho, Nishikyogoku, Ukyo-ku
Kyoto 615-0882, Japan
jdoi@mail.koka.ac.jp

Zhenxing Pan
Doctoral candidate
Graduate School of Engineering
Osaka University
2-1 Yamadaoka, Suita
Osaka 565-0871, Japan
pan@is.eei.eng.osaka-u.ac.jp

Kunichika Tsumoto
Specially Appointed Assistant Professor
The Center for Advanced Medical
　Engineering and Informatics
Osaka University
2-2 Yamadaoka, Suita
Osaka 565-0871, Japan
tsumoto@pharma2.med.osaka-u.ac.jp

ISBN 978-4-431-53861-5　　　　ISBN 978-4-431-53862-2 (eBook)
DOI 10.1007/978-4-431-53862-2
Springer Tokyo Dordrecht Heidelberg London New York

Library of Congress Control Number: 2009943556

MATLAB® is a registered trademark of The MathWorks, Inc., 3 Apple Hill Drive, Natick, MA 01760-2098, USA. http://www.mathworks.com

Cover Illustration: © MEIcenter

Printed on acid-free paper

Springer is part of Springer Science+Business Media (www.springer.com)

Preface

Life is a dynamic, ambiguous, and fleeting system. Computational biology and systems biology use mathematical or computational models, usually denoted as dynamical systems such as difference equations or differential equations, for the analysis and understanding of such biological systems. In general, these models include many parameters, and these parameters inherently possess much ambiguity or uncertainty; thus it is important to treat the models with the consideration of such parameter uncertainty or changes. Bifurcation theory in nonlinear dynamical systems provides us with a powerful tool for the analysis of the effect of parameter change on a system and detects a critical parameter value when the qualitative nature of the system changes, whereas numerical or computer simulations give us only one solution under a fixed set of parameter values and initial values.

Our aim is to provide an introduction to computational electrophysiology, particularly to the nonlinear dynamics of Hodgkin–Huxley (HH)-type models, together with a brief introduction to the dynamical system and to bifurcation analysis. This textbook includes many examples of numerical computations of bifurcation analysis on various models, ranging from the classical HH model of a squid giant axon to the Luo–Rudy (LRd) dynamic model of a cardiac cell, using the famous computer software XPPAUT [see B. Ermentrout (2002) *Simulating, Analyzing, and Animating Dynamical Systems: A Guide to XPPAUT for Researchers and Students.* SIAM, Philadelphia], including AUTO software (http://cmvl.cs.concordia.ca/auto/). We hope that this work will be useful as a practical, quick guide to the numerical bifurcation analysis of other models that readers have to analyze.

This textbook, Volume 2 in the three-volume series *A First Course in "In Silico Medicine,"* is organized as follows:

Chapter 1 provides the basics of dynamical system theory necessary for understanding the dynamics and bifurcations arising in the models encountered in the succeeding chapters. Chapter 1 also provides very brief instructions in XPPAUT, so that readers can perform bifurcation analysis similar to that presented in this textbook.

Chapter 2 explores the nonlinear dynamics of the famous HH model of a neuronal excitable membrane in detail and explains what makes up its neuronal features, such as the excitability threshold and refractoriness. Part of the explanation of HH dynamics is indebted to the very old but impressive paper [FitzHugh (1960)

Thresholds and plateaus in the Hodgkin-Huxley nerve equations, J. Gen. Physiol. 43, pp. 867–896]. This seminal work provides a good example of the direction that contemporary computational biology should take. The geometric methods in the dynamical system rather than brute-force computer simulations are very useful for understanding the essential features of neuronal and nonlinear dynamics.

Using the Bonhoeffer–van der Pol (BVP) neuronal model [or the FitzHugh–Nagumo (FHN) model], which is a simplification of the HH model, Chapter 3 explains neuronal dynamics more geometrically and intuitively. In the rest of Chapter 3, we take a slightly lengthy, roundabout route via more simplified or abstract neuronal models. In particular, we show that many neuronal models arising in different contexts are systematically explained by a simple one-dimensional mapping called the phase transition curve (PTC). Through these excursions, we would like to emphasize the importance of not only the physiological or physical models but also the abstract models.

Chapter 4 returns to the HH model but explores it from a different viewpoint: robustness and sensitivity. The bifurcation structure of the HH model on various bifurcation parameters is thoroughly analyzed, and thus the robustness and sensitivity of the HH model on various parameters are clarified.

Chapter 5 analyzes the bifurcation structure of other HH-type models: the Yanagihara–Noma–Irisawa (YNI) model of a cardiac pacemaker cell and the LRd model of a cardiac ventricular cell. These models are HH-type models but possess greater complexity than the classical HH model of a squid giant axon. Thus the robustness and sensitivity of cardiac cells on various parameters (ion channels) are also explored.

We would like to express our sincere gratitude to many people, especially to Professors Shunsuke Sato, Hiroshi Kawakami, and Jose Pedro (Pepe) Segundo for their stimulating discussion, teaching, and encouragement. We are very grateful to Professors Taishin Nomura and Yoshihisa Kurachi, and also for the support of the Global COE Program "In Silico Medicine" at Osaka University.

Masao Tanaka
Series Editor

Shinji Doi
Junko Inoue
Zhenxing Pan
Kunichika Tsumoto
Authors

Contents

Chapter 1
A Very Short Trip on Dynamical Systems

The theory of dynamical systems has grown up to a very big discipline in mathematics now, and deals mainly with *measurable* (or *ergodic*), *topological*, and *differentiable* (or *smooth*) dynamics of systems. The dynamical system theory is related to other areas of mathematics also, for example, to the (analytic or geometric) singular perturbation theory. The singular perturbation theory is very important to analyze and understand the mathematical models in the electrophysiology (see Coombes and Bressloff 2005; Fenichel 1979; Guckenheimer 1996; Izhikevich 2006; Jones 1996; Jones and Kopell 1994, for example), but is not treated in this book. This chapter briefly introduces only the minimal materials in the broad dynamical system theory, necessary for the understanding of the following mathematical and computational models in the electrophysiology. This chapter also introduces the practical method of numerical bifurcation analysis by using the numerical analysis softwares XPPAUT and AUTO.

1.1 Difference Equations, Maps, and Linear Algebra

Natural, social and artificial systems change hour by hour. Dynamical system is a mathematics for the modeling and the analysis of such systems' behavior. Dynamical systems incorporate the *state* and its time change in a system. Consider a difference equation or discrete-time dynamical system:

$$x(n + 1) = f(x(n)), \quad x(n) \in \mathcal{R}^N, \quad n = 0, 1, 2, \dots. \tag{1.1}$$

$x(n)$ is an N-dimensional real vector (i.e. $x(n) = (x_1(n), x_2(n), \dots, x_N(n))^T$ where $(\cdot)^T$ denotes a transpose of a vector or a matrix $(\cdot)$) and represents a state (called a *state point* or simply a *state*) of a system at a time n. The N-dimensional map f transfers the state $x(n)$ into another state $x(n + 1) = f(x(n))$. The state $x(0)$ is referred as the *initial state*. We can denote the n-th state $x(n)$ using the initial state:

$$x(n) = f^n(x(0)) = \overbrace{f \circ f \circ \cdots \circ f}^{n \text{ times}}(x(0)). \tag{1.2}$$

S. Doi et al., *Computational Electrophysiology*,
DOI 10.1007/978-4-431-53862-2_1, © Springer 2010

Iterations of the map f generate a sequence of states (or a set of states):

$$\{x(n)\} = x(0), x(1), x(2), \ldots \tag{1.3}$$

which is called an *orbit* (or *trajectory* or *solution*) of the system (1.1). A set of all states (in our case, $\mathcal{R}^N$) is called a *state space* (or phase space). If a state $x^* \in \mathcal{R}^N$ satisfies the relation $f(x^*) = x^*$, then the orbit started from x^* stays at the point x^* forever:

$$x(n) = x(n-1) = \cdots = x(1) = x(0) = x^*. \tag{1.4}$$

The special state x^* is called a *fixed point* or an *equilibrium point*.

The simplest example of the discrete dynamical system (1.1) is the case that the map f is a linear matrix A:

$$x(n+1) = Ax(n), \quad x(n) \in \mathcal{R}^N, \quad n = 0, 1, 2, \ldots, \tag{1.5}$$

where A is an invertible $N \times N$ matrix. Note that the origin $x = (0, \ldots, 0)^T$ is always a unique fixed point since this system is linear and A is invertible. Then, the orbit or general solution of this system is obtained as

$$x(n) = A^n x(0). \tag{1.6}$$

First, consider a simple case that $N = 2$ and A is a 2×2 matrix:

$$x(n+1) = \begin{pmatrix} a & b \\ c & d \end{pmatrix} x(n), \qquad x(n) = \begin{pmatrix} x_1(n) \\ x_2(n) \end{pmatrix}, \qquad n = 0, 1, 2, \ldots. \tag{1.7}$$

We also suppose that $b = c = 0$. In this case, we can solve (obtain its orbit) easily:

$$x(n) = A^n x(0) = \begin{pmatrix} a^n & 0 \\ 0 & d^n \end{pmatrix} x(0) = \begin{pmatrix} a^n x_1(0) \\ d^n x_2(0) \end{pmatrix}. \tag{1.8}$$

Figure 1.1 shows various orbits for various initial values $(x_1(0), x_2(0))^T$. The panel (a) corresponds to the case $0 < a < d < 1$ (and $b = c = 0$). All orbit move toward the origin $(0, 0)^T$ as $n \to \infty$. This result is apparent since $|a|, |d| < 1$. (Note that the orbits are discrete points [dots]. The lines with arrow between dots are not orbits, which are shown for the sake of presentation clarity in the figure.) Orbits with special initial values on the x or y axes move straightly toward the origin. General orbits do not move straightly but approach the origin *along* the y-axis. This is because $|a| < |d|$ and thus the x-component vanishes more quickly than the y-component; the y-component is *dominant* when the time n is large enough. In this case, the origin is the fixed point of this system and also is *stable* since all orbits nearby the origin approach it as time n increases. Another example is illustrated in panel (b) in which $0 < a < 1 < d$ and $b = c = 0$. In this case, all orbits on the x-axis converge on the origin since $|a| < 1$. On the other hand, all orbits on the y-axis diverge since $|d| > 1$. Other orbits except for these special orbits diverge upward or downward although they approach the y-axis (the x-component vanishes). In this case, the origin is an *unstable* fixed point.

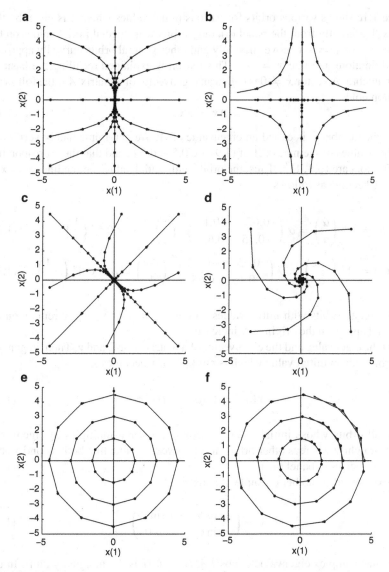

Fig. 1.1 Phase plane and orbits of the linear dynamical system (1.7). (**a**) $a = 0.5$, $b = 0$, $c = 0$, $d = 0.8$. (**b**) $a = 0.7$, $b = 0$, $c = 0$, $d = 1.5$. (**c**) $a = 0.65$, $b = -0.15$, $c = -0.15$, $d = 0.65$. (**d**) $a = 0.7 \cos(\pi/6)$, $b = -0.7 \sin(\pi/6)$, $c = -b$, $d = a$. (**e**) $a = 1 \cos(\pi/6)$, $b = -1 \sin(\pi/6)$, $c = -b$, $d = a$. (**f**) $a = 1 \cos(0.51)$, $b = -1 \sin(0.51)$, $c = -b$, $d = a$. The origin $(0, 0)$ is the fixed point and specifically called (**a**) *node*, (**b**) *saddle*, (**c**) *node*, (**d**) *focus*, (**e, f**) *center*

Next, consider more general case of the matrix A:

$$A = \begin{pmatrix} 0.65 & -0.15 \\ -0.15 & 0.65 \end{pmatrix}. \tag{1.9}$$

Figure 1.1c shows various orbits for various initial values when A is given by this matrix. Differently from the panel (a), only orbits with special initial values on the line $y = x$ or $y = -x$ move straightly and other general orbits curvedly approach the origin along the line $y = -x$. This reason is explained recalling the basic of linear algebra. A vector $x \neq 0$ is called an eigenvector of a matrix A if the following equation holds

$$Ax = \lambda x, \tag{1.10}$$

where the number λ is called an eigenvalue of A. We can immediately verify that the eigenvalues of the matrix A of (1.9) are 0.5 and 0.8, and that the corresponding eigenvectors are $(\pm 1, 1)^T$. Thus, the orbits with initial values on the lines $y = x$ or $y = -x$ become as follows:

$$x(n) = A^n \begin{pmatrix} \alpha \\ \alpha \end{pmatrix} = \alpha \begin{pmatrix} 0.65 & -0.15 \\ -0.15 & 0.65 \end{pmatrix}^n \begin{pmatrix} 1 \\ 1 \end{pmatrix} = \alpha \times 0.5^n \begin{pmatrix} 1 \\ 1 \end{pmatrix}, \tag{1.11}$$

$$x(n) = A^n \begin{pmatrix} -\alpha \\ \alpha \end{pmatrix} = \alpha \begin{pmatrix} 0.65 & -0.15 \\ -0.15 & 0.65 \end{pmatrix}^n \begin{pmatrix} -1 \\ 1 \end{pmatrix} = \alpha \times 0.8^n \begin{pmatrix} -1 \\ 1 \end{pmatrix}. \tag{1.12}$$

Therefore, all orbits with initial values on lines $y = x$ or $y = -x$ remain on the lines and approach the origin as n increases.

Let the eigenvalue and the eigenvector of a matrix A be λ and v. Then, in general, the orbit with an initial value which is multiple of v becomes

$$x(n) = A^n(\alpha v) = \alpha \lambda^n v. \tag{1.13}$$

Thus, all orbits whose initial values (vectors) are scalar-multiples of eigenvector move straightly whereas other orbits move curvedly. (Note that both x- and y-axes are eigenvectors in panel (a).)

Next, consider another example of a matrix A:

$$A = \begin{pmatrix} r \cos \theta & -r \sin \theta \\ r \sin \theta & r \cos \theta \end{pmatrix} \tag{1.14}$$

which has complex eigenvalues $r \cos \theta \pm i r \sin \theta$ (i is the imaginary unit). In this case, the orbit is obtained as follows:

$$x(n) = A^n x(0) = \begin{pmatrix} r \cos \theta & -r \sin \theta \\ r \sin \theta & r \cos \theta \end{pmatrix}^n x(0)$$

$$= r^n \begin{pmatrix} \cos n\theta & -\sin n\theta \\ \sin n\theta & \cos n\theta \end{pmatrix} x(0). \tag{1.15}$$

Thus, the orbit moves counter-clockwise at the speed proportional to θ and approaches the origin if $0 \leq r < 1$. Figure 1.1d shows various orbits when $r = 0.7$ and $\theta = \pi/6$. All orbits wind counter-clockwise and approach the origin since

$r < 1$. If $r > 1$ the directions of all orbits will be reversed and all orbits spiral outward the phase plane. Panels (e) and (f) explain a special case between $r < 1$ and $r > 1$: $r = 1$. In panel (e), we set $\theta = \pi/6$, and all orbits started from the x-axis (the $x > 0$ range) return to their starting points after twelve iterations of the map: $x(12) = x(0)$. This is apparent because $n\theta = 12(\pi/6) = 2\pi$ in (1.15). In general, an orbit $\{x(n)\}$ such that $x(p) = x(0)$ and $x(i) \neq x(0)$ for $1 \leq i < p$ is called a periodic orbit with period p. All orbits in panel (e) are periodic orbits with period 12. In panel (f), the value of θ is set as $\theta = 0.51$ which is slightly smaller than $\theta = \pi/6 \approx 0.53$ of panel (e). Although, the orbits in panels (e) and (f) look similar each other, the orbits of (f) can not reach the x-axis after twelve iterations of the map since $\theta = 0.51 < \pi/6$. Note that those orbits in (f) cannot become periodic since any integer-multiples of 0.51 (rational number) cannot become multiple of 2π (irrational number). Such orbits in panel (f) are called *quasi-periodic* orbits. (Compare with the similar phase plane of Fig. 1.3 in the following continuous-time case.)

Exercise 1.1. The following is an example of MATLAB commands to make the phase plane of Fig. 1.1d:

```
%This is comment. Phase plane of 2D difference eqs.
clf; mm=5; axis([-mm, mm, -mm, mm]); axis('square');
xlabel('x(1)');ylabel('x(2)');
hold on
plot([-mm mm], [0 0], 'k'); plot([0 0], [-mm mm], 'k');
a=0.7*cos(pi/6); b=-0.7*sin(pi/6); c=-b; d=a;
A=[a b; c d];
x0=[3.5;3.5];
for i=1:15; plot(x0(1),x0(2),'.k'); x=A*x0;
  quiver(x0(1),x0(2),x(1)-x0(1),x(2)-x0(2),0,'k'); x0=x;
end
x0=[-3.5;3.5];
for i=1:15; plot(x0(1),x0(2),'.k'); x=A*x0;
  quiver(x0(1),x0(2),x(1)-x0(1),x(2)-x0(2),0,'k'); x0=x;
end
x0=[-3.5;-3.5];
for i=1:15; plot(x0(1),x0(2),'.k'); x=A*x0;
  quiver(x0(1),x0(2),x(1)-x0(1),x(2)-x0(2),0,'k'); x0=x;
end
x0=[3.5;-3.5];
for i=1:15; plot(x0(1),x0(2),'.k'); x=A*x0;
  quiver(x0(1),x0(2),x(1)-x0(1),x(2)-x0(2),0,'k'); x0=x;
end
hold off
```

Execute these commands on MATLAB and confirm that Fig. 1.1d can be obtained. Modify these commands to make other phase planes of Fig. 1.1.

(The MATLAB command `quiver(x,y,u,v)` plots vector(s) as arrow(s) with components (u,v) at the point(s) (x,y). For more explanation, we can use MATLAB help as: `help quiver`. The option '.k' of `plot(x0(1),x0(2),'.k')` means that it plots with a dot mark and with black color.) □

1.2 Differential Equations, Vector Fields, and Phase Planes

In this section, we consider the case that time is continuous. (Please compare the following description with that of the previous section.) In this case, dynamical systems are usually described by differential equations rather than difference equations. The dynamical system which corresponds to the discrete-time dynamical system (1.1) becomes

$$\frac{d}{dt}x(t) = f(x(t)), \quad x(t) \in \mathcal{R}^N, \quad t \in \mathcal{R}. \tag{1.16}$$

The solution $x(t)$ which satisfies this differential equation is called an *orbit* or *trajectory* of the system (1.16) similarly to the discrete-time system. The state $x(0)$ is referred as the initial state again. The special state point x^* such that $f(x^*) = 0$ is called a *fixed point*, an *equilibrium point* or a *steady state* similarly to the discrete-time case (but note that the difference of the definition between discrete- and continuous-time cases: $f(x^*) = x^*$ vs. $f(x^*) = 0$). The right-hand side (r.h.s.) $f(x(t)) \in \mathcal{R}^N$ of the differential equation (1.16) is a vector and is called a *vector field*. The vector field assigns the vector $f(x)$ to each point x of the state space $\mathcal{R}^N$.

The simplest example of the continuous dynamical system (1.16) is also the case that the map f is a linear matrix A:

$$\frac{d}{dt}x(t) = Ax(t), \quad x(t) \in \mathcal{R}^N, \quad t \in \mathcal{R}, \tag{1.17}$$

where A is an invertible $N \times N$ matrix. Note that only the origin $x = (0, \ldots, 0)^T$ is the fixed point or an equilibrium point since this system is linear and A is invertible. The general solutions can be obtained by

$$x(t) = \exp(At)x(0), \tag{1.18}$$

where the *exponential* of a matrix A is formally defined as follows:

$$\exp(A) \equiv I + \frac{1}{1!}A + \frac{1}{2!}A^2 + \frac{1}{3!}A^3 + \cdots = \sum_{k=0}^{\infty} \frac{1}{k!}A^k. \tag{1.19}$$

Note that the actual computation of this exponential for a general matrix A is not so easy.

First, consider a finite *difference* approximation of the differential $dx(t)/dt$ in (1.17):

$$\frac{d}{dt}x(t) \approx \frac{x(t + \delta t) - x(t)}{\delta t}, \tag{1.20}$$

where δt is a small time interval. Then (1.17) becomes

$$\frac{x(t + \delta t) - x(t)}{\delta t} = f(x(t)) = Ax(t) \tag{1.21}$$

and thus we obtain

$$x(t + \delta t) = x(t) + \delta t\, f(x(t)) = x(t) + \delta t A x(t) = (I + \delta t A)x(t). \quad (1.22)$$

Namely, the discrete dynamical system (1.5) has appeared again, although the time step is not unity but δt here: this is not important. Thus, the orbit of (1.22) is obtained by (if $t/\delta t$ is an integer)

$$x(t) = (I + \delta t A)^{t/\delta t} x(0). \quad (1.23)$$

Let us consider some examples. We suppose that $N = 2$ and A is a 2×2 matrix:

$$\frac{d}{dt}x(t) = \begin{pmatrix} a & b \\ c & d \end{pmatrix} x(t), \quad x(t) = \begin{pmatrix} x_1(t) \\ x_2(t) \end{pmatrix}, \quad t \in \mathcal{R}. \quad (1.24)$$

Figure 1.2 shows such examples where we set as $a = -0.5$, $b = 0$, $c = 0$ and $d = -0.2$. The curve with dots shows the orbit where $\delta t = 1$ and the broken curve without dot corresponds to $\delta t = 0.5$. The solid curve without dot shows the theoretical solution:

$$x(t) = \exp(At)x(0) = \exp\left\{ \begin{pmatrix} -0.5t & 0 \\ 0 & -0.2t \end{pmatrix} \right\} x(0)$$

$$= \begin{pmatrix} e^{-0.5t} & 0 \\ 0 & e^{-0.2t} \end{pmatrix} x(0). \quad (1.25)$$

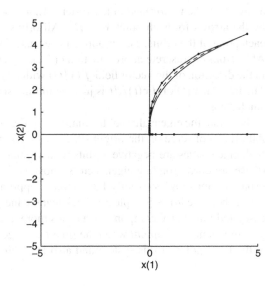

Fig. 1.2 Orbits of a finite difference approximation (1.22) of linear differential equations. $a = -0.5$, $b = 0$, $c = 0$, $d = -0.2$. The *curve with dots* shows the orbit where $\delta t = 1$ and the *broken curve without dot* corresponds to $\delta t = 0.5$. The *solid curve without dot* shows the theoretical solution

Orbits are shown for two initial values: $x(0)^T = (4.5, 4.5)$ and $(4.5, 0)$. For the former initial value, we can see that the discretely approximated orbit approaches the theoretical one as the value of δt decreases. Note that the approximate orbit with $\delta t = 1$ is not so bad, although the value of δt is extremely large and thus the orbit jumps at big step (see the dot marks). For the latter initial value, two approximate orbits and the theoretical one overlap completely. We also note that for $\delta t = 1$, the discrete approximation (1.23) becomes

$$x(t) = (I + \delta t A)^{t/\delta t} x(0) = \begin{pmatrix} 1 - 0.5\delta t & 0 \\ 0 & 1 - 0.2\delta t \end{pmatrix}^{t/\delta t} x(0)$$

$$= \begin{pmatrix} 0.5 & 0 \\ 0 & 0.8 \end{pmatrix}^t x(0). \tag{1.26}$$

In the original dynamical system (differential equations), since the eigenvalues of the matrix A are -0.5 and -0.2, and are less than zero, thus the fixed point or equilibrium point of the differential equation (1.24) is stable (please try to take a limit of the r.h.s. of (1.25) when $t \to \infty$). In the approximated dynamical system (1.26), since the eigenvalues of the matrix of the right-hand side (r.h.s.) are 0.5 and 0.8, and their absolute values are less than unity, thus the fixed point at the origin of the discretely approximated system is also stable when $\delta t = 1$. However, if the value of δt is increased over 4, the fixed point becomes unstable in the discrete approximation since the absolute value of one eigenvalue $(1 - 0.5\delta t)$ exceeds unity.

Let us return to the linear differential equations (1.24) apart from the discrete approximations. Figure 1.3 shows various orbits for various initial values $(x_1(0), x_2(0))^T$. Panel (a) corresponds to the case $a < d < 0$ (and $b = c = 0$). The (red) heavy curves are the orbits with various initial values and short lines with arrow show the *vector field* of the system (1.24); the vectors $f(x) = Ax$ are shown by the arrows for each point $x \in \mathcal{R}^2$. All orbits are tangent to the vector field at each point of the orbits. This is apparent because $dx(t)/dt$ means a *velocity* vector. Also, from the discrete approximation (1.22), we can see that the orbit $x(t)$ moves in the direction of the vector field $f(x(t))$ with an *infinitesimally* small step size δt. The integral $x(t)$ of $dx(t)/dt$ is just the infinite summation of such infinitesimally small steps.

Note that the eigenvalues of the matrix A in Fig. 1.3a are $a = -0.5$ and $d = -0.2$. All orbits converge on the origin $(0, 0)^T$ as $t \to \infty$. This result is apparent since both eigenvalues are negative. Orbits with special initial values on the x or y axes (these are corresponding eigenvectors) move straightly toward the origin. General orbits do not move straightly but curvedly approach the origin *along* the y-axis. This is because $|a| > |d|$ (please check that the inequality is opposite in the discrete case) and thus the x-component vanishes more quickly than the y-component; the y-component is *dominant* when the time t is large enough. In this case, the origin is the fixed point of this system and also is *stable*. Another example is illustrated

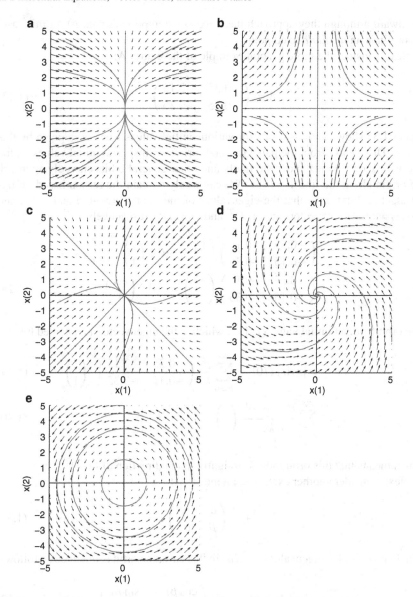

Fig. 1.3 Vector fields and orbits of the linear dynamical system (1.24) in a two-dimensional phase plane. (a) $a = -0.5$, $b = 0$, $c = 0$, $d = -0.2$. (b) $a = -0.3$, $b = 0$, $c = 0$, $d = 0.5$. (c) $a = -0.35$, $b = -0.15$, $c = -0.15$, $d = -0.35$. (d) $a = -0.3$, $b = -0.5$, $c = -b$, $d = a$. (e) $a = 0.0$, $b = -2\pi$, $c = -b$, $d = a$. The origin $(0, 0)$ is the fixed point and specifically called (a) *node*, (b) *saddle*, (c) *node*, (d) *focus*, (e) *center*

in panel (b) in which $a < 0 < d$ and $b = c = 0$. In this case, all orbits on the x-axis converge on the origin since $a < 0$. On the other hand, all orbits on the y-axis diverge since $d > 0$. Other orbits except for these special orbits diverge upward or downward although they approach the y-axis (x-components vanish). In this case, the origin is an *unstable* fixed point.

Next, consider the more general example of the matrix A:

$$A = \begin{pmatrix} -0.35 & -0.15 \\ -0.15 & -0.35 \end{pmatrix}. \tag{1.27}$$

Figure 1.3c shows various orbits for various initial values when A is given by this matrix. Differently from the panel (a), only orbits with special initial values on the line $y = x$ or $y = -x$ move straightly and other general orbits curvedly approach the origin along the line $y = -x$. We can easily explain the reason based on linear algebra. First, note that the eigenvalues of the matrix are -0.5 and -0.2, and corresponding eigenvectors are $(1, 1)^T$ and $(-1, 1)^T$, respectively:

$$\begin{pmatrix} -0.35 & -0.15 \\ -0.15 & -0.35 \end{pmatrix} \begin{pmatrix} 1 \\ 1 \end{pmatrix} = -0.5 \begin{pmatrix} 1 \\ 1 \end{pmatrix},$$

$$\begin{pmatrix} -0.35 & -0.15 \\ -0.15 & -0.35 \end{pmatrix} \begin{pmatrix} -1 \\ 1 \end{pmatrix} = -0.2 \begin{pmatrix} -1 \\ 1 \end{pmatrix}. \tag{1.28}$$

The orbit from an initial value $(1, 1)^T$, which is the eigenvector, is obtained by

$$x(t) = \exp(At)x(0) = \sum_{k=0}^{\infty} \frac{t^k}{k!} \begin{pmatrix} -0.35 & -0.15 \\ -0.15 & -0.35 \end{pmatrix}^k \begin{pmatrix} 1 \\ 1 \end{pmatrix} \tag{1.29}$$

$$= \sum_{k=0}^{\infty} \frac{t^k(-0.5)^k}{k!} \begin{pmatrix} 1 \\ 1 \end{pmatrix} = e^{-0.5t} \begin{pmatrix} 1 \\ 1 \end{pmatrix} \tag{1.30}$$

which means that this orbit moves straightly in the direction of $(1, 1)^T$.

Next, consider another example of a matrix A:

$$A = \begin{pmatrix} \alpha & -\beta \\ \beta & \alpha \end{pmatrix} \tag{1.31}$$

which has complex eigenvalues $\alpha \pm i\beta$. In this case, the orbit is obtained as follows:

$$x(t) = \exp(At)x(0) = e^{\alpha t} \begin{pmatrix} \cos \beta t & -\sin \beta t \\ \sin \beta t & \cos \beta t \end{pmatrix} x(0). \tag{1.32}$$

Thus, when $\beta > 0$, the orbit moves counter-clockwise at a speed proportional to β and approaches the origin if $\alpha < 0$. Figure 1.3d shows various orbits when $\alpha = -0.3$ and $\beta = 0.5$. All orbits wind counter-clockwise toward the origin. Panel (e) illustrates a special case: the $\alpha = 0$ case. All orbits do not approach the origin but return to the initial state points. Since $\beta = 2\pi$, all orbits in (e) are periodic orbits with period one. In panel (e), orbits $\{x(t), 0 \leq t \leq 0.98\}$ are shown.

Exercise 1.2. The following is an example of MATLAB commands to draw the phase plane of Fig. 1.3d:

```
%This is comment. Phase plane of 2D differential eqs.
clf; mm=5; axis([-mm, mm, -mm, mm]); axis('square');
xlabel('x(1)');ylabel('x(2)');
hold on
plot([-mm mm], [0 0], 'k'); plot([0 0], [-mm mm], 'k');
global a b c d;
r=5;X=[-r:r/10:r];Y=[-r:r/10:r];
a=-0.3;b=-0.5; c=-b; d=a;
[XX,YY]=meshgrid(X,Y); F=( a*XX+b*YY );G=( c*XX+d*YY );
quiver(X,Y,F,G,'k');
t0=0;tf=100;
x0=[3.5;3.5]; [t,x]=ode23('linFlow',[t0, tf], x0');
plot(x(:,1),x(:,2),'-r');
x0=[-3.5;3.5]; [t,x]=ode23('linFlow',[t0, tf], x0');
plot(x(:,1),x(:,2),'-r');
x0=[-3.5;-3.5]; [t,x]=ode23('linFlow',[t0, tf], x0');
plot(x(:,1),x(:,2),'-r');
x0=[3.5;-3.5]; [t,x]=ode23('linFlow',[t0, tf], x0');
plot(x(:,1),x(:,2),'-r');
hold off
```

The MATLAB command ode23 numerically solves the differential equations which are described in the linFlow.m file:

```
%This is linFlow.m file.
function xdot=linFlow(t,x)
 global a b c d
 xdot=zeros(2,1);
 xdot(1)=a*x(1)+b*x(2);
 xdot(2)=c*x(1)+d*x(2);
```

Execute these commands on MATLAB and confirm that Fig. 1.3d can be obtained. Modify these commands to make other phase planes of Fig. 1.3. ∎

1.3 Linearization, Stabilities, Coordinate Transformation

Although we have already explained about the stabilities of fixed points (equilibrium points), let us summarize about the stabilities here. Consider the discrete-time linear dynamical system (1.5). In the system, the origin $\mathbf{0}$ is the fixed point and is stable (i.e. $x(n) \to \mathbf{0}$, $n \to \infty$) if absolute values of all eigenvalues of A are less than unity. In the case of the continuous-time dynamical system (1.17), the equilibrium point at the origin is stable if the real parts of all eigenvalues of A are less than zero (negative).

1.3.1 Linearization and Stabilities

Next, consider nonlinear dynamical systems. Let x^* be an equilibrium point of the discrete-time dynamical system (1.1) or the continuous-time dynamical system (1.16): $f(x^*) = x^*$ in (1.1) or $f(x^*) = 0$ in (1.16). The Taylor expansion of the function $f(x)$ near the equilibrium point x^* are obtained as follows:

$$f(x) = f(x^*) + \mathcal{D}f(x^*)(x - x^*) + \mathcal{O}(||x - x^*||^2), \qquad (1.33)$$

where $\mathcal{O}(||x - x^*||^2)$ denotes the higher-order terms (second-order terms and higher terms) and $\mathcal{D}f(x^*)$ is the Jacobian matrix:

$$\mathcal{D}f(x^*) = \left(\begin{array}{ccc} \dfrac{\partial f_1(x)}{\partial x_1} & \dfrac{\partial f_1(x)}{\partial x_2} & \cdots \\ \dfrac{\partial f_2(x)}{\partial x_1} & \dfrac{\partial f_2(x)}{\partial x_2} & \cdots \\ \vdots & \vdots & \cdots \end{array} \right)_{x=x^*}, \qquad (1.34)$$

$$x = (x_1, x_2, \cdots)^T, \quad f(x) = (f_1(x_1, x_2, \cdots), f_2(x_1, x_2, \cdots), \cdots)^T.$$

Near the equilibrium point x^*, we can neglect (under some conditions) the higher-order terms $\mathcal{O}(||x - x^*||^2)$ since $\mathcal{O}(||x - x^*||^2)$ becomes small when $||x - x^*||$ is small. Then, we can obtain a *linearized system* or *linearization* of (1.1) and (1.16), respectively as follows:

$$z(n + 1) = Az(n), \quad A = \mathcal{D}f(x^*), \qquad (1.35)$$

$$\frac{d}{dt}z(t) = Az(t), \quad A = \mathcal{D}f(x^*), \qquad (1.36)$$

where we have made use of the change of a variable $z(n) = x(n) - x^*$ (or $z(t) = x(t) - x^*$).

1.3.2 Coordinate Transformations

In dynamical systems, a coordinate transformation or a change of variables is very useful and often simplify our view about the dynamical systems. In this subsection, we briefly explain such coordinate transformations.

Consider the linear discrete-time system (1.7). If we change the state variable x to z by $x = Tz$ where T is an invertible (or regular) $N \times N$ matrix, the linear discrete-time dynamical system (1.5) becomes

$$z(n + 1) = T^{-1}x(n + 1) = T^{-1}Ax(n) = (T^{-1}AT)z(n). \qquad (1.37)$$

Thus, if the matrix $(T^{-1}AT)$ has a simple form such as diagonal matrices, the coordinate transformation makes the analysis of the original system much simpler. For the linear continuous-time dynamical system (1.17), the transformed system can be similarly obtained as

$$\frac{d}{dt}z(t) = T^{-1}\frac{d}{dt}x(t) = T^{-1}Ax(t) = (T^{-1}AT)z(t). \tag{1.38}$$

In the case of *nonlinear* dynamical systems (the discrete-time system (1.1) or continuous-time system (1.16)), we can treat the coordinate transformation similarly. Let us consider the case that the coordinate transformation is also nonlinear. Let $x = g(z)$ be such a nonlinear transformation with an inverse g^{-1}, where x denotes $x(n)$ or $x(t)$ (similarly z denotes $z(n)$ or $z(t)$). Then the transformations of (1.1) and (1.16) are respectively obtained as

$$z(n+1) = g^{-1}(x(n+1)) = (g^{-1} \circ f)(x(n)) = (g^{-1} \circ f \circ g)(z(n)) \tag{1.39}$$

and

$$\frac{d}{dt}z(t) = g^{-1}\left(\frac{d}{dt}x(t)\right) = (g^{-1} \circ f)(x(t)) = (g^{-1} \circ f \circ g)(z(t)), \tag{1.40}$$

where the symbol "$\circ$" denotes a composition of maps.

1.4 Nonlinear Dynamical Systems and Bifurcations

Most examples of (discrete- or continuous-time) dynamical systems treated in previous sections are linear. In this section, we proceed to more general dynamical systems: nonlinear dynamical systems, and discuss the *bifurcation* of a certain dynamical system. Roughly speaking, the word bifurcation means the change of the number and/or stability of equilibrium points or periodic orbits.

1.4.1 Bifurcations of Discrete Dynamical Systems

Lift, Degree and Rotation Number

Let us consider the simplest case ($N = 1$ case) of the discrete-time dynamical system (mapping) (1.1); consider the following *one-dimensional map* or *1D map*:

$$x(n+1) = f(x(n)), \; x(n) \in \mathcal{S} \quad (n = 0, 1, 2, \ldots), \tag{1.41}$$

where S is the unit circle (i.e. an interval $[0, 1]$ identified unity with zero) and the map $f(x)$ is the function from S to S with a certain smoothness (differentiability). Such a map is often called a *circle map*.

We say that x^* is a periodic point of period n if $f^n(x^*) = x^*$ and $f^i(x^*) \neq x^*$ for $1 \leq i < n$. The set $\{f^i(x^*), \ 0 \leq i < n\}$ of periodic points is the periodic orbit. The periodic point x^* with unit period ($f(x^*) = x^*$) is called the fixed point particularly. The fixed point x^* is *asymptotically stable* if there exists δ and for any $x \in (x^* - \delta, x^* + \delta)$, $f^n(x) \to x^*$ as $n \to \infty$. The asymptotic stability of a periodic point or periodic orbit can be defined similarly. If $|f'(x^*)| < 1$ (the prime means d/dx), the fixed point x^* is asymptotically stable while it is unstable if $|f'(x^*)| > 1$. When $|f'(x^*)| = 1$, the stability is neutral and is not determined by the coefficient $|f'(x^*)|$ of a linear term, and nonlinear terms or higher order terms of the Taylor expansion of $f(x)$ are necessary for the determination of the stability. Similarly, if the following relation:

$$\left| \frac{d}{dx} f^n(x^*) \right| = \left| f' \left(f^{n-1}(x^*) \right) \right| \cdot \left| f' \left(f^{n-2}(x^*) \right) \right| \cdots \left| f' \left(f(x^*) \right) \right| \cdot \left| f' \left(x^* \right) \right| < 1 \tag{1.42}$$

holds for a periodic point x^*, the periodic point x^* and periodic orbit $\{x^*, f(x^*), \ldots, f^{n-1}(x^*)\}$ are asymptotically stable. When $df^n(x^*)/dx = 0$, the periodic orbit is the most stable and is called *super stable* especially.

Next, we suppose that $f : S \to S$ is continuous. A continuous map $F : \mathcal{R} \to \mathcal{R}$ is a *lift* of f if and only if:

$$F(x) \bmod 1 = f(x \bmod 1), \quad x \in \mathcal{R} \tag{1.43}$$

where "mod" means a modulus. Note that there are countably infinite lifts F for each f. There exists an integer d such that $F(x+1) = F(x) + d$ for all real numbers x. This number d is called the *degree* of F or of f. If F is a non-decreasing function, the limit

$$\rho = \lim_{n \to \infty} \frac{F^n(x)}{n} \tag{1.44}$$

can be well defined and is called the *rotation number* of F. Then, the number (ρ mod 1) is also called the rotation number of f. The rotation number ρ is very useful to characterize the dynamics of one-dimensional maps. For example, f has a periodic orbit if and only if ρ is a rational number. Also, if F is non-decreasing and $\rho = p/q$ with p and q being relatively prime integers, the periods of all periodic orbit of f become q.

Figure 1.4 shows examples of linear maps on S:

$$x(n+1) = f(x(n)) \equiv x(n) + a \bmod 1, \quad x(n) \in S \quad (n = 0, 1, 2, \ldots) \tag{1.45}$$

which is a simple shift of $x(n)$ by a in a sense of modulo unity. Panel (a) is the case of $a = 0.4$. Thick lines denote the graph of $f(x)$ and thin lines show the orbit

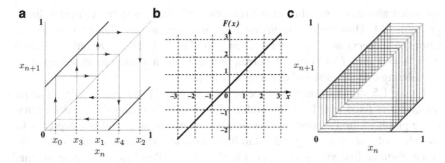

Fig. 1.4 Examples of the linear map (1.45) and its orbit. (**a**) Periodic case, $a = 0.4$. *Thick lines* are the graph of map and *thin lines* show an example of a periodic orbit with period five. The *diagonal thin line* is the line $x(n + 1) = x(n)$ using which orbits of a map can be graphically obtained. (**b**) Lift of (a). The map of (a) defined on S is extended to the one on the whole real axis $\mathcal{R}$. (**c**) Quasi-periodic case, $a = \pi/10$. Note that the notation x_n is used instead of $x(n)$, in this figure

of the map (1.45): $x(0) = 0.1$, $x(1) = 0.5$, $x(2) = 0.9$, $x(3) = 0.3$, $x(4) = 0.7$, $x(5) = 0.1 = x(0)$ which is a periodic orbit with period five. Diagonal line $x(n + 1) = x(n)$ is also shown. It can be easily shown that all initial values $x \in S$ have a period five. These periodic orbits or points are neither asymptotically stable nor unstable since the derivative $f'(x) \equiv 1$. Panel (b) shows the continuous lift $F(x)$ of $f(x)$ in panel (a), from which we can see that the map of (a) is of degree 1 and has a rotation number $0.4 = 2/5$ since

$$\lim_{n\to\infty} \frac{F^n(x)}{n} = \lim_{n\to\infty} \frac{x + n \times 0.4}{n} = 0.4 = \frac{2}{5}.$$

This result agrees with the result that the periods of all periodic orbits of (a) are five. Panel (c) is the case of $a = \pi/10$ whose rotation number is an irrational number $\pi/10$ and thus the map of (c) does not possess any periodic orbits: any orbits show *quasi-periodic* behavior and never return to the previously visited points. Note, however, that digital computers can not realize an irrational number in a strict sense and thus computer generated orbits may show periodic behavior with a long period.

Bifurcations of the Standard Circle Map

Next let us examine how the solution structure of a certain map is changed with a continuous change of a parameter of the map, i.e. we study the *bifurcation* structure of a typical circle map called the *standard circle* or *sine-circle* map:

$$x(n + 1) = f(x(n)) \equiv x(n) + a + b \sin(2\pi x(n)) \mod 1$$
$$x(n) \in S \quad (n = 0, 1, 2, \ldots). \tag{1.46}$$

As stated above, the stability of a fixed point x^* changes at the critical value of $|f'(x^*)| = 1$. Thus, roughly speaking, a certain bifurcation (a change of qualitative nature) may occur at a fixed point when $|f'(x^*)| = 1$. There are only two cases $f'(x^*) = 1$ and $f'(x^*) = -1$ corresponding to such a bifurcation since we are considering one-dimensional dynamical systems.

Figure 1.5 shows examples of the sine-circle map (1.46) and its orbit. The parameter a is fixed to 0.1 and the value of b is slightly changed. The thick curve is the graph of the function $f(x)$ which is continuous on S and has degree 1 for all values of b shown in Fig. 1.5. Thin lines show an orbit with an initial value 0.1 and a diagonal line $x(n + 1) = x(n)$ is also shown. Panel (a) is the case of small b value in which there is neither fixed point nor stable periodic orbit and the map shows a so-called quasi-periodic behavior. If b is slightly increased ($b = 0.1$), the graph of $f(x)$ is tangent to the diagonal line at $x = 3/4$ where a *saddle-node* or *fold* bifurcation occurs. After the fold bifurcation occurrence, a pair of fixed points (stable one $\bar{x}$ and unstable one $\hat{x}$) are generated (see Fig. 1.5c). If the value of b is further increased ($b \approx 0.337$), the slope at the stable fixed point $\bar{x}$ becomes -1 and a *period-doubling* or *flip* bifurcation occurs. After this bifurcation, the fixed point $\bar{x}$ becomes unstable and a stable periodic orbit with period two (the thick square of Fig. 1.5e) appears (i.e. the unit period of the fixed point is *doubled* to two), and any orbits asymptotically approach this periodic orbit.

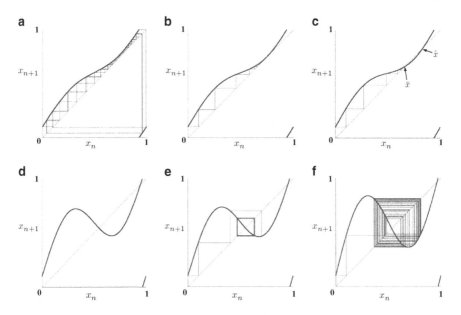

Fig. 1.5 Examples of the sin-circle map (1.46). $a = 0.1$. (**a**) $b = 0.08$. (**b**) $b = 0.1$. (**c**) $b = 0.12$. (**d**) $b = 0.31$. (**e**) $b = 0.35$. (**f**) $b = 0.46$. Note that the notation x_n is used instead of $x(n)$, in this figure

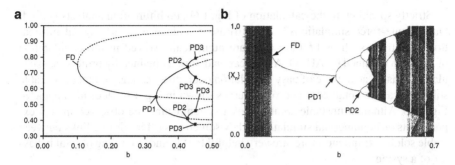

Fig. 1.6 One-parameter bifurcation diagram of the sine-circle map for the bifurcation parameter b when $a = 0.1$. (**a**) Bifurcation diagram obtained by AUTO (Doedel et al. 1995). (**b**) Bifurcation diagram obtained by brute-force computer simulations of (1.46): Asymptotic sequences $\{x(n)\}$, ($n = 201, \ldots, 700$) produced by (1.46) were plotted for each of 1,000 equally spaced b values on the interval $[0, 0.7]$

These results are summarized in the *bifurcation diagram* of Fig. 1.6a where the abscissa (the horizontal axis) is *the bifurcation parameter b* and the ordinate (the vertical axis) shows the x values of fixed points or periodic orbits. Namely, the bifurcation diagram shows how the (stationary or asymptotic) solutions of the system (1.46) change as the parameter b varies. The solid curves show the stable fixed points or periodic points and the broken curves unstable ones. At the fold bifurcation point FD ($b = 0.1$) of Fig. 1.6a, a pair of stable and unstable fixed points are simultaneously generated. This stable fixed point undergoes a period-doubling bifurcation at the point PD1 ($b \approx 0.337$) after which the fixed point becomes unstable and a stable periodic orbit with period two is generated. This periodic orbit undergoes a subsequent period-doubling bifurcation at PD2 ($b \approx 0.427$) and a further period-doubled periodic orbit is generated. A further period-doubling bifurcation occurs at PD3 although a period-eight orbit is not shown in the figure. In fact, such a cascade of period-doubling bifurcations continues infinitely and it is very difficult to trace out all such bifurcations. Figure 1.6a is computed using the bifurcation analysis software called AUTO (Doedel et al. 1995) which is very useful to analyze the nonlinear dynamical systems described by difference or differential equations.

Figure 1.6b is also a bifurcation diagram calculated by the computer simulation of (1.46): Asymptotic sequences $\{x(n)\}$, ($n = 201, \ldots, 700$) produced by (1.46) were plotted for each of 1,000 equally spaced b values on the interval $[0, 0.7]$. In the range of $b < 0.1$, (1.46) shows a quasi-periodic behavior (many dots are plotted). A cascade of period-doubling bifurcations explained in (a) is also seen. In the right of the period-doubling cascade, chaotic solutions appear (deep-black region). Figure 1.5f shows an example of such a chaotic solution. Although there exist many deep-black regions in Fig. 1.6b, only black regions in $b > 0.1$ corresponds to chaotic solutions. Roughly speaking, as the portion of $f(x)$ with $|f'(x)| > 1$ is increased, the map $f(x)$ tends to create chaotic solutions.

Strictly speaking, in the calculation of Fig. 1.6b, no bifurcation analysis is used: Only "brute-force simulations" of (1.46) are examined whereas several bifurcation conditions such as $|f'(x)| = 1$ are numerically solved in the computation of Fig. 1.6a obtained by AUTO. A bifurcation diagram obtained by brute-force simulations has an advantage of easy calculation but a disadvantage in that only stable solutions can be obtained; the broken curves shown in Fig. 1.6a does not appear in Fig. 1.6b. Although unstable solutions of a system cannot be observed in real experiments or in numerical simulations, they sometimes offer missing links between stable solutions, and therefore, are very helpful for the understanding of total behavior of a system.

Exercise 1.3. Compute numerically the bifurcation diagram for the logistic map $f(x) = ax(1-x)$ with a as a bifurcation parameter. Next, compute the fixed points of the map analytically (as a function of the parameter a). What are the stabilities of these fixed points? Also, compute the periodic orbits with period two analytically (A *nonlinear* coordinate transformation (variable change) from x to z such that $z = a(1 - 2x)/2$ may be useful for such analytical calculations).   □

1.4.2 Bifurcations of Continuous Dynamical Systems

Saddle-Node Bifurcation

Consider a (nonlinear) one-dimensional differential equation:

$$\dot{x} = f_\lambda(x), \quad f_\lambda(x) = -x^2 + \lambda, \quad x, \lambda \in \mathcal{R}, \tag{1.47}$$

where $\dot{x}$ denotes the time derivative: $\dot{x} = dx/dt$, and λ is a parameter. Figure 1.7 shows the graph of $f_\lambda(x)$ as a function of x for various values of λ. Panel (c) corresponds to the case $\lambda = 1$. In this case, the graph of the function $f_\lambda(x)$ intersects with the x-axis at the two points: $x = \pm 1$. These points are the equilibrium points

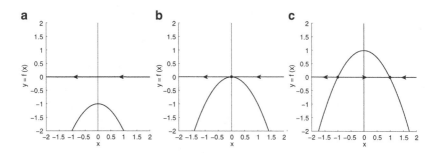

Fig. 1.7 Vector fields and orbits of the one-dimensional *nonlinear* dynamical system (1.47): $\dot{x} = -x^2 + \lambda$. (**a**) $\lambda = -1$. (**b**) $\lambda = 0$. (**c**) $\lambda = 1$

of the system (1.47) since $f_{\lambda=1}(\pm 1) = 0$. In the figure, the x-axis corresponds to the *one-dimensional* state space and the equilibrium points are shown by big dots. The arrows show the direction of the *one-dimensional* vector field or orbits on the x-axis. For example, in the range $x > 1$ of Fig. 1.7c, the direction is left since $f_\lambda(x) < 0$. In the range $-1 < x < 1$, the direction becomes right, and thus all orbits started from the range $-1 < x$ move toward the equilibrium point at $x = 1$. Therefore, $x = 1$ is a *stable* equilibrium point. On the other hand, $x = -1$ is a unstable equilibrium point, and all orbits started from $x < -1$ diverge leftward.

In Fig. 1.7b, the value of λ is decreased to zero. Thus, the graph of the function $f_\lambda(x)$ is tangent to the x-axis at the unique point: $x = 0$. There is only one equilibrium point. All orbits started in the range $x > 0$ approach $x = 0$, whereas all orbits started from the range $x < 0$ move leftward and finally diverge. If the value of λ is further decreased, the graph of $f_\lambda(x)$ never cross the x-axis: there is no equilibrium point. All orbits move leftward and diverge. As the value of the parameter λ changes, the behavior of the dynamical system (1.47) changes, particularly the number and/or the stability of equilibrium points change. This is a bifurcation similarly to the discrete-time systems. $\lambda = 0$ is the bifurcation point and this bifurcation is called a saddle-node (SN) or a fold bifurcation again (compare this with the saddle-node bifurcation in the discrete-time systems).

Figure 1.8 is the bifurcation diagram of the dynamical system (1.47). The abscissa denotes the bifurcation parameter λ and the ordinate the x-value; the equilibrium points of the system (1.47) are plotted as a function of λ. The solid and broken curves denote the stable and unstable equilibria, respectively. At the point labeled SN, a saddle-node bifurcation occurs and a pair of stable and unstable equilibria are generated.

Next, we consider the following two-dimensional system:

$$\begin{pmatrix} \dot{x}_1 \\ \dot{x}_2 \end{pmatrix} = \begin{pmatrix} -x_1{}^2 + \lambda \\ -x_2 \end{pmatrix}. \tag{1.48}$$

It is apparent that this system has two equilibrium points $(\pm\sqrt{\lambda}, 0)$ if $\lambda > 0$. Figure 1.9 shows the state plane (phase plane) of this system. When $\lambda < 0$ (panel a),

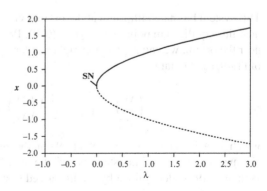

Fig. 1.8 Bifurcation diagram of the one-dimensional nonlinear dynamical system (1.47): $\dot{x} = -x^2 + \lambda$

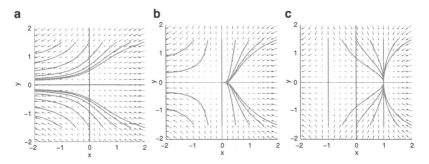

Fig. 1.9 Vector fields and orbits of the two-dimensional system (1.48). (**a**) $\lambda = -1$. (**b**) $\lambda = 0$. (**c**) $\lambda = 1$

all orbits diverge leftward approaching the horizontal axis. When $\lambda > 0$ (panel c), orbits with an initial value $x_1(0) > -\sqrt{\lambda}$ converge on the right equilibrium point $(+\sqrt{\lambda}, 0)$ and orbits with $x_1(0) < -\sqrt{\lambda}$ diverge leftward approaching the horizontal axis. The right equilibrium point is stable and is called a node, whereas the left equilibrium point $(-\sqrt{\lambda}, 0)$ is called a saddle. A saddle is a point that it is stable in a one direction (vertical, in this case) and is unstable in another direction (horizontal). When $\lambda = 0$ (panel b), both saddle and node collide. Therefore, this bifurcation is called a saddle-node bifurcation. The bifurcation diagram is completely the same as that of the one-dimensional system (1.47). This is apparent since the variable x_2 does not affect the dynamics of x_1 in the two-dimensional system (1.48). Although the example (1.48) looks very artificial, the essential dynamics of general saddle-node bifurcations is one-dimensional similarly to (1.48).

Hopf Bifurcation

Consider the following two-dimensional nonlinear dynamical system:

$$\begin{pmatrix} \dot{x}_1 \\ \dot{x}_2 \end{pmatrix} = \begin{pmatrix} \lambda & -1 \\ 1 & \lambda \end{pmatrix} \begin{pmatrix} x_1 \\ x_2 \end{pmatrix} - \begin{pmatrix} x_1(x_1^2 + x_2^2) \\ x_2(x_1^2 + x_2^2) \end{pmatrix}. \qquad (1.49)$$

This system has an equilibrium at the origin $(x_1, x_2) = (0, 0)$. Consider the system near this equilibrium point: $(x_1, x_2) \approx (0, 0)$. Because $|x_1| \ll |x_1 \cdot x_1^2|$ and so on near the origin, we can neglect the higher-order terms (the second term of the r.h.s. of (1.49)) and obtain

$$\begin{pmatrix} \dot{x}_1 \\ \dot{x}_2 \end{pmatrix} = \begin{pmatrix} \lambda & -1 \\ 1 & \lambda \end{pmatrix} \begin{pmatrix} x_1 \\ x_2 \end{pmatrix} \equiv A \begin{pmatrix} x_1 \\ x_2 \end{pmatrix} \qquad (1.50)$$

which is a linearized system of (1.49) at the origin. Clearly the matrix A has eigenvalues $\lambda \pm i$. The stability of the equilibrium $(0, 0)$ of the nonlinear system (1.49) is determined by the linearized system (1.50). Thus, the equilibrium

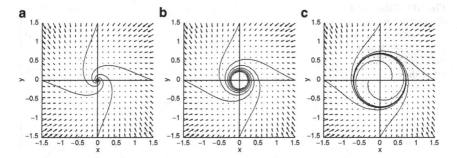

Fig. 1.10 Vector fields and orbits of the two-dimensional dynamical system (1.49) which illustrates the super-critical *Hopf bifurcation*. (**a**) $\lambda = -0.5$. (**b**) $\lambda = 0$. (**c**) $\lambda = 0.5$

point is stable or unstable if $\lambda < 0$ or $\lambda > 0$, respectively. When $\lambda = 0$, the stability cannot be determined by the linearized system but it is determined by the nonlinear (higher) terms of the system (1.49). Later explanation will show that the equilibrium point is still stable even when $\lambda = 0$.

Figure 1.10 shows the vector field and orbits of the nonlinear system (1.49) when (a) $\lambda = -0.5$, (b) $\lambda = 0$, (c) $\lambda = 0.5$. Orbits in panel (a) move spirally toward the origin; we can see that the origin is a stable equilibrium point. On the other hand, when $\lambda = 0.5$ (see panel c), the origin becomes unstable. Orbits near the origin leave the origin and move toward a certain *circle* with a radius of $1/\sqrt{2}$. Orbits far from the origin also move toward the circle. This circle itself is a special orbit (periodic orbit) called a *limit cycle*. Panel (b) illustrates a special case between panels (a) and (c): $\lambda = 0$. In the figure, we can see a circle, but it is not really a limit cycle. The origin is still stable although the stability is weak. All orbits in panel (b) will finally approach the origin if we prolong the simulation time.

To summarize, the equilibrium point at the origin changes its stability as the value of λ changes its sign from negative to positive. When the equilibrium becomes unstable, a stable limit cycle appears. This is called a *Hopf bifurcation*. Figure 1.11 is the bifurcation diagram which summarizes the dependency of the dynamics of the system (1.49) on the parameter λ. The abscissa shows the bifurcation parameter λ and the ordinate the variable x_1 of the system (1.49). The solid and broken curves (lines) show the x_1 component of the stable and unstable equilibrium points as a function of λ, respectively. The filled circles show the maximum values of x_1 of *stable* periodic orbits. At the point labeled by HB, the Hopf bifurcation occurs and the stability of the equilibrium point is changed and the stable periodic orbit is generated (bifurcated).

Consider a similar system to (1.49):

$$\begin{pmatrix} \dot{x}_1 \\ \dot{x}_2 \end{pmatrix} = \begin{pmatrix} \lambda & -1 \\ 1 & \lambda \end{pmatrix} \begin{pmatrix} x_1 \\ x_2 \end{pmatrix} + \begin{pmatrix} x_1(x_1{}^2 + x_2{}^2) \\ x_2(x_1{}^2 + x_2{}^2) \end{pmatrix}, \qquad (1.51)$$

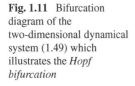

Fig. 1.11 Bifurcation diagram of the two-dimensional dynamical system (1.49) which illustrates the *Hopf bifurcation*

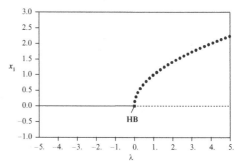

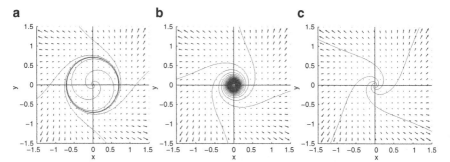

Fig. 1.12 Vector fields and orbits of the two-dimensional dynamical system (1.51) which illustrates the *sub-critical* Hopf bifurcation. (**a**) $\lambda = -0.5$. (**b**) $\lambda = 0$. (**c**) $\lambda = 0.5$

where only the sign of the nonlinear term is different from (1.49). Figure 1.12 shows the vector field and orbits of this system (1.51) when (a) $\lambda = -0.5$, (b) $\lambda = 0$, (c) $\lambda = 0.5$. First, note that the stability of the equilibrium point $(0, 0)$ in both cases (a) and (c) is the same as that of Fig. 1.10 since the stability is determined by the linear terms. The stability in (b), however, differs between Figs. 1.10 and 1.12 since the nonlinear terms are different. Figure 1.12c is very similar to Fig. 1.10a. Note that, however, the direction of orbits is different (the stability of the equilibrium is different), and the orbits move outward and diverge. In Fig. 1.12a, the equilibrium point is stable and thus orbits near the origin approach the equilibrium. The orbits far from the origin move outward and diverge finally. There is a boundary between the inward and outward orbits. This boundary is a circle with a radius $\sqrt{2}$ and a special orbit called a limit cycle also. This limit cycle, differently from Fig. 1.10c, is *unstable* and nearby orbits do not approach it. When $\lambda = 0$, the stability of the equilibrium is not determined by the linear terms but by nonlinear terms, and it is unstable as is clarified later. Thus all orbits nearby the origin moves outward and finally diverge, although the speed of divergence is very slow (compare panels b and c in Fig. 1.12).

Figure 1.13 is the bifurcation diagram of (1.51). The open circles show *unstable* periodic orbits. The stability of the equilibrium points of Fig. 1.13 is the same as those of Fig. 1.11 (except for the $\lambda = 0$ case). The way of generation of periodic

Fig. 1.13 Bifurcation diagram of the two-dimensional dynamical system (1.51) which illustrates the *sub-critical* Hopf bifurcation

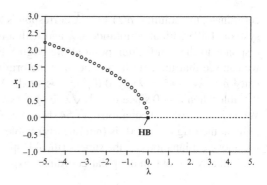

orbits, however, is different. Namely, in Fig. 1.13, the periodic orbits are unstable and the region of λ where the periodic orbits exist is the range *below* the bifurcation point $\lambda = 0$, whereas the region in Fig. 1.11 is the range *above* the Hopf bifurcation. Therefore, the Hopf bifurcations in Figs. 1.11 and 1.13 are called *super-* and *sub-critical* Hopf bifurcations, respectively. (Note that the words super and sub do not really denote the direction of the branch of periodic orbits, but do denote the stability of periodic orbits; there exists a sub-critical Hopf bifurcation that unstable periodic orbits exist *above* the sub-critical Hopf bifurcation point. For example, if we change λ by $-\lambda$ in (1.51), then the bifurcation diagram becomes the similar one to Fig. 1.13 where the right and left are reversed.)

In order to understand the dynamics of the system (1.49), it is very useful to use a coordinate transformation, particularly to use a complex variable. Consider a coordinate transformation from the two-dimensional variable (x_1, x_2) to the one-dimensional *complex* variable z: $z = x_1 + ix_2$. Of course, we denote the complex conjugate as $\bar{z} = x_1 - ix_2$ and the absolute value as $|z|^2 = x_1{}^2 + x_2{}^2$. Then, we obtain

$$
\begin{aligned}
\dot{z} &= \dot{x}_1 + i\dot{x}_2 \\
&= \lambda x_1 - x_2 - x_1(x_1{}^2 + x_2{}^2) + i(x_1 + \lambda x_2) - ix_2(x_1{}^2 + x_2{}^2) \\
&= \lambda(x_1 + ix_2) + i(x_1 + ix_2) - (x_1 + ix_2)(x_1{}^2 + x_2{}^2) \\
&= \lambda z + iz - z|z|^2.
\end{aligned}
\tag{1.52}
$$

Furthermore, using the *polar coordinate* $z = \rho e^{i\phi}$, we have

$$
\dot{z} = \dot{\rho}e^{i\phi} + i\rho e^{i\phi}\dot{\phi} = \lambda\rho e^{i\phi} + i\rho e^{i\phi} - \rho e^{i\phi}\rho^2.
\tag{1.53}
$$

Comparing both sides of the second equality of this equation, we obtain

$$
\begin{aligned}
\dot{\rho} &= \lambda\rho - \rho^3 = \rho(\lambda - \rho^2), \\
\dot{\phi} &= 1.
\end{aligned}
\tag{1.54}
$$

Thus, we can see that $\dot{\rho} = 0$ for $\rho = \sqrt{\lambda}$ if $\lambda \geq 0$ (Note that ρ is the *amplitude* and thus $\rho = -\sqrt{\lambda}$ is not allowed). Since the (ρ, ϕ) coordinate system is the polar coordinate, the solution $\rho(t) = \sqrt{\lambda}$ corresponds to a *periodic orbit* of the original system (1.49) with an amplitude $\sqrt{\lambda}$ and with a unit angular velocity ($\rho = 0$ corresponds to the equilibrium point at the origin). From the first equation of (1.54), we can see that the equilibrium point at the origin ($\rho = 0$) is stable when $\lambda < 0$ since $\dot{\rho} = \lambda\epsilon - \epsilon^3 \approx \lambda\epsilon < 0$ if $\rho = \epsilon$ is small. Also, the periodic orbit ($\rho = \sqrt{\lambda}$) is stable when $\lambda > 0$ since $\dot{\rho} = \lambda(\sqrt{\lambda} \pm \epsilon) - (\sqrt{\lambda} \pm \epsilon)^3 \approx \mp 2\lambda\epsilon$ if $\rho = \sqrt{\lambda} \pm \epsilon$. When $\lambda = 0$, the first equation of (1.54) becomes $\dot{\rho} = -\rho^3$ and thus the equilibrium point at the origin ($\rho = 0$) is (nonlinearly) stable. These results are consistent with the above explanations on the super-critical Hopf bifurcation (Fig. 1.11).

Similarly, we obtain the polar-coordinate expression for the system (1.51):

$$\dot{\rho} = \lambda\rho + \rho^3 = \rho(\lambda + \rho^2),$$
$$\dot{\phi} = 1. \tag{1.55}$$

Only the difference from (1.54) is the sign of the nonlinear (cubic) term in the first equation. Thus, this system has an *unstable* periodic orbit with an amplitude $\sqrt{-\lambda}$ when $\lambda < 0$. Also, when $\lambda = 0$, the equilibrium point at the origin ($\rho = 0$) is (nonlinearly) unstable since the first equation of (1.55) becomes $\dot{\rho} = +\rho^3$. See the bifurcation diagram of Fig. 1.13.

Exercise 1.4. Use the polar coordinate transformation $x_1 = \rho\cos\phi$, $x_2 = \rho\sin\phi$ for (1.49) to directly obtain the same result as (1.54).   ☐

1.4.3 Forced Dynamical Systems and Poincaré Maps

One-dimensional map of (1.41) is seemingly too simple to model and analyze real systems. The one-dimensional map, however, arises commonly from general n-dimensional dynamical systems. Consider the following "forced" dynamical system (differential equations):

$$\frac{d\boldsymbol{x}}{dt} = \boldsymbol{f}(\boldsymbol{x}) + \boldsymbol{g}(t), \quad \boldsymbol{x} \in \mathcal{R}^n, \quad \boldsymbol{f}: \mathcal{R}^n \to \mathcal{R}^n, \quad \boldsymbol{g}: \mathcal{R} \to \mathcal{R}^n, \tag{1.56}$$

where $\boldsymbol{g}$ denotes an external input (a forcing term) applied to the system and is a time-periodic function with period T: $\boldsymbol{g}(t + T) = \boldsymbol{g}(t)$. Suppose that we make a sampling (or an observation) of the state variable $\boldsymbol{x}(t)$ of (1.56) at every time interval T, and let us denote the state $\boldsymbol{x}(nT)$ at the n-th observation by $\boldsymbol{x}_n$. Also, let us denote the solution of (1.56) with an initial state $\boldsymbol{x}(0) = \boldsymbol{x}_0$ after a time t as $\Phi(\boldsymbol{x}_0, t)$. Then, the relation between $\boldsymbol{x}_n$ and $\boldsymbol{x}_{n+1}$ is denoted by the following discrete-time dynamical system:

$$\boldsymbol{x}_{n+1} = F(\boldsymbol{x}_n) \equiv \Phi(\boldsymbol{x}_n, T), \quad F: \mathcal{R}^n \to \mathcal{R}^n \quad (n = 0, 1, 2, \ldots). \tag{1.57}$$

Usually, the dynamical system (1.56) without the external forcing ($g(t) \equiv 0$) has an *attractor* (asymptotically stable set) and the state point $x(t)$ asymptotically moves near the attractor. Even in the case that the forcing term $g(t)$ is present, if $g(t)$ is small enough relatively to the stability of the attractor or $g(t)$ is impulsive like a neuronal action potential, the state point $x(t)$ can be assumed to move near the attractor. Then the map F is (approximately) a map from the attractor to itself. If the dynamical system (1.56) with $g(t) \equiv 0$ shows an oscillatory or periodic behavior, the attractor is a one-dimensional closed orbit called a *limit cycle*. Then the n-dimensional map F can be reduced to a one-dimensional map on the limit cycle. (Even if the system is not oscillatory but excitable like a neuron, F can become one-dimensional. See Doi and Sato 1995.) By a suitable choice of a coordinate (usually called a phase) on the limit cycle, the map becomes the one-dimensional circle map (1.41).

1.4.4 Attractors and Basins of Attraction

So far, we have seen that dynamical systems can have various stable orbits such as stable equilibrium points, stable limit cycles. Such a stable set or a geometric object which consists of state points is called an *attractor*, since the set attracts nearby orbits. Dynamical systems can have multiple attractors: two equilibrium points, an equilibrium point and a limit cycle, and so on. Then, the final fate of an orbit depends on its initial state. (Note that linear [discrete- or continuous-time] dynamical system have only one equilibrium or fixed point, thus only nonlinear dynamical systems can have multiple attractors.)

Consider a ball (considered to be a point particle) moving under the influence of friction in a *one-dimensional* potential or a slope $V(x)$ as depicted in the upper panel of Fig. 1.14 (Mcdonald et al. 1985). There are two points x_1^* and x_2^* where the potential takes its minimal value. Thus, the ball eventually settles down to one of two points. Since the movement of the ball is determined by its initial position x and velocity dx/dt, the state space (phase plane) is two-dimensional as shown in the lower panel. There are two stable equilibrium points denoted by filled circles in the phase plane. There is also an unstable equilibrium point denoted by an open circle, which corresponds to the potential peak in the center of the upper panel. If an initial state is taken from the white region depicted in the x–dx/dt phase plane, the orbit starting from the initial state eventually converges to the left equilibrium point x_2^* while the orbits from shaded (yellow) region come to rest at the right equilibrium x_1^*. Each region (a set of state points) which finally leads to an attractor (a stable equilibrium point in this case) is called a *basin of attraction* or a *basin* simply. These two basins are interweaved with each other, and thus a small change in the ball's position and/or velocity leads a big difference in its final fate. The basins in this simple example are not actually so interweaved. In fact, the shape of basins can become more complicated like *fractal* (Mcdonald et al. 1985). Anyway, we should pay much attention

Fig. 1.14 A simple ball
moving in a one-dimensional
potential illustrating a
complicated shape of a basin
of attraction. This is modified
from Mcdonald et al. (1985)

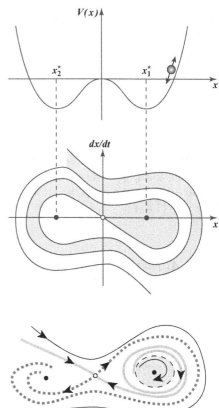

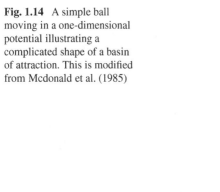

Fig. 1.15 Example of multiple attractors and basins often encountered in neuronal models

on the shape of basins particularly in numerical simulations, since arbitrary choice
of an initial value leads to different simulation results.

   Figure 1.15 schematically illustrates typical examples of multiple attractors and
basins often encountered in neuronal models. There are two stable equilibrium
points (filled circles) and one unstable equilibrium (open circle) in both left and
right phase planes. The dynamics, however, are much different in the two phase
planes. The center unstable equilibrium is a saddle; there are two special orbits
(thick solid curves with light-blue color) converge to the unstable equilibrium point
and other two special orbits (thick dotted curves with red color) converge to the
point "when time is reversed." Such special orbits of thick solid (light-blue) and
thick dotted (red) are called *stable and unstable manifold*, respectively. Although
these orbits are special in that we should carefully choose an initial state in order to
find those orbits in numerical simulations, they play a very important role in order to
clarify the dynamics of dynamical systems. The shaded (yellow) region is the basin
of attraction of the right stable equilibrium in both phase planes. In the left figure,
the basin is separated by the stable manifolds of the saddle. (Note that orbits started
outside the basin approach the "left" equilibrium even if their initial states are closer

to the "right" equilibrium.) On the other hand, in the right phase plane of Fig. 1.15, the basin of the right equilibrium is small and is separated by a closed curve (broken curve). The closed circle is an unstable periodic orbit or limit cycle. In such a phase plane, there occurs an interesting phenomenon; orbits started closely the unstable limit cycle (but outside the limit cycle) will wind the limit cycle many times and finally converge to the left stable equilibrium point.

In summary, the clarification of the geometric structure of such special (unstable) orbits as stable and unstable manifolds, unstable limit cycles, etc., is very important for the understanding of the dynamics of systems.

1.5 Computational Bifurcation Analysis

In this section, we explain how to compute numerically the solutions and bifurcation diagrams of dynamical systems. First, note that usual *numerical* bifurcation analyses do not use the direct numerical simulations of dynamical systems (differential or difference equations), but find the solutions of such dynamical systems by different ways. Numerical bifurcation analyses trace continuously the solutions started from a certain known solution (an equilibrium point or a periodic orbit) if such solutions are "connected." Thus, the numerical bifurcation analyses can find not only stable solutions but also unstable solutions. Also, such bifurcation analyses can be performed in relatively short computational time since they do not require lengthy numerical simulations of dynamical systems.

There are several computer softwares for the numerical bifurcation analyses, such as CONTENT (http://www.enm.bris.ac.uk/staff/hinke/dss/continuation/content.html), MATCONT (http://www.matcont.ugent.be/), DDE-BIFTOOL (http://twr.cs.kuleuven.be/research/software/delay/ddebiftool.shtml), Oscil8 (http://oscill8.sourceforge.net/), BunKi (http://bunki.ait.tokushima-u.ac.jp:50080/), and so on. The most famous computer software of bifurcation analysis is AUTO (AUTO97 and AUTO07P) and many bifurcation diagrams in this book were made by using AUTO (http://cmvl.cs.concordia.ca/auto/). In this section, however, let us explain another software called XPPAUT or XPP (XPP-Aut: X-Windows Phase Planes plus Auto) which adds the user-friendly graphical user interface (GUI) and several useful functions to AUTO. XPPAUT can be obtained from http://www.math.pitt.edu/~bard/xpp/xpp.html. Note that the following examples of XPPAUT were performed on both MacOS X and Windows XP. Some appearances and functions of XPPAUT may be different between the operating systems. The usage of XPPAUT, however, is essentially the same and does not depend on the operating systems.

In order to start the XPPAUT on MacOS with X-windows installed, just type xppaut. On the Windows machine, start cygwin and type xinit or startx to start the X-window, and type ./xpp.bat in the xppall folder (under the usual installation of the XPPAUT).

When we start XPPAUT by typing xppaut, the small window shown in Fig. 1.16 will appear. We can see a list of files (in this case, only one file: hopf.ode) therein.

Fig. 1.16 Starting window of
XPPAUT (or simply XPP).
We can choose an ode file to
be analyzed

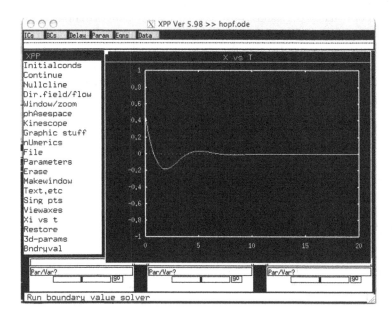

Fig. 1.17 The main window of the XPPAUT. On the *top row* and in the *left column*, there are many
buttons. The *right big panel* is the main graphic window

We can scroll up and down the list (if this includes many files) by using ^^ (or ^)
and vv (or v), and change directories. If we click a mouse button on the hopf.ode,
The big window shown in Fig. 1.17 comes up. This is the main window of XPPAUT
(note that at the start-up, such a drawing of a solution in the "Graphic Window"
which is the right panel in the main window does not appear). In order to exit the
XPPAUT, click on the File button in the left column of the XPPAUT main window
and then click on Quit.

The content of the hopf.ode file is as follows:

```
# Filename: hopf.ode
# This is comment. Example of the Hopf bifurcation
# in the differential equations
x'= L*x - y - x*( x^2 + y^2 )
y'= x + L*y - y*( x^2 + y^2 )
# parameters
par L=-0.5
# initial conditions
init x=0.5, y=0.5
done
```

Fourth and fifth lines describe the differential equations which are the same as (1.49). Parameters and initial conditions are also described in the file.

Solving Differential Equations

In the leftmost of the top row of Fig. 1.17, we can see the ICs button where we can set initial values. If we click on the button, a small Initial Data window shown in Fig. 1.18 appears. We can see the initial "initial values" described in the hopf.ode file have already set to the X and Y variables. If we click on the Go button in the Initial Data window, a solution started from the initial condition will be drawn as in Fig. 1.17 where the abscissa and the ordinate are the time T and the variable X, respectively. If we change the initial values in the Initial Data window and press Go button again, another solution started from a new initial condition will be drawn (superimposed).

We can change the X vs. T plot (which is shown on the top of the "Graphic Window" in the main window) to the Y vs. X plot (i.e. X-Y phase plane) as follows. Click on the Viewaxes button and then 2D button in the left column of the XPP main window, the 2D View window such as Fig. 1.19 comes up (note that Fig. 1.19 is the window after some changes were made). Then, change *X-axis and *Y-axis to X and Y, respectively. We can change the range of X (Y) axis by changing Xmin and Xmax (Ymin and Ymax, resp.).

Clicking on the Initialconds button and then (M)ouse (or m(I)ce) allows us to set initial values by using a mouse. By m(I)ce, we can draw many orbits and only one orbit by (M)ouse. Figure 1.20 shows the X-Y phase plane and several

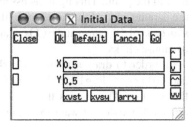

Fig. 1.18 The Initial Data window in which we can set initial values or initial states of differential equations

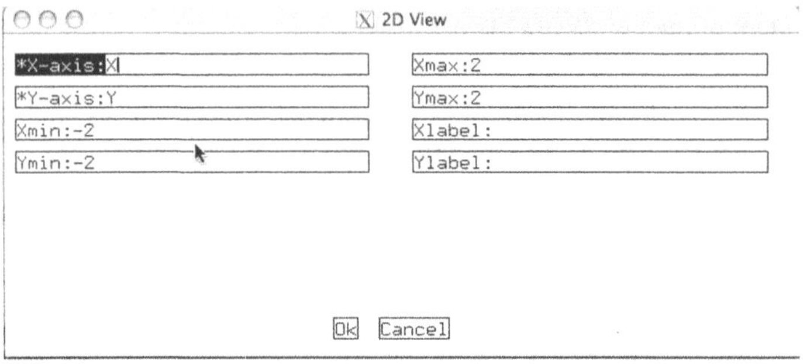

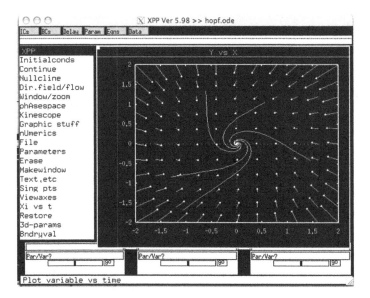

Fig. 1.19 The 2Dview window. We can change the variables and their range to be plotted in the X- and Y-axes

Fig. 1.20 The vector field and several orbits (solutions) of differential equations

orbits started at several initial values inputted by mouse clicks (note that the vector field is also drawn by arrows. We will explain how to draw such a vector field later.) We can see that the origin $(0, 0)$ is a stable equilibrium point and all orbits spirally tend to the point. This figure is essentially the same as Fig. 1.10 (check the value of the parameter L of the hopf.ode file). Note that the (M) and (I) are the *keyboard shortcuts* of (M)ouse and m(I)ce, and we can type these keys on a keyboard instead of mouse clicks.

In order to draw a vector field shown in Fig. 1.20, click on Dir.field/flow and (D)irect Field. Then a message "Grid: 10" appears in the second line of the window. This means the number of grids where the vector field is drawn by arrows. If we do not want to change this default value, just hit return (or enter) key on a keyboard. Then, a vector field will appear. We can draw various orbits superimposed on this vector field by the method explained above. In order to clear the graphic window, just click a mouse button on `Erase`.

Exercise 1.5. Change the value of the parameter as $L = 0$ and $L = +0.5$, and draw several obits with different initial values and vector fields. Then obtain similar $X-Y$ phase plane to Fig. 1.10b,c. (In order to change the parameter value, use the `Param` button on the top row of the XPP main window.)   □

Bifurcation Diagrams

Next, let us draw a bifurcation diagram using XPPAUT. First, we note that the numerical bifurcation analysis performs a *continuation*. Namely, changing parameter values step by step, it seeks various solutions (equilibrium points and periodic orbits) from an *initial solution*. (Note that an initial solution does not mean an initial condition.) Thus, at first, we must choose or fix an initial solution from which the continuation is made. Often, we use a stable equilibrium point as an initial solution. To obtain such a stable equilibrium point, use `Initialconds` and `(G)o`, and then draw a solution, which approaches the stable equilibrium point, from any initial value. Next, repeat both `Initialconds` and `(L)ast` (I and L keys) several times. This command solves or integrates differential equations from the final (last) value of previous solution. Thus we can obtain the approximate values of the stable equilibrium point. We can see these values by clicking on `Sing pts` and `(G)o`.

File and Auto buttons create a new window such as Fig. 1.21 for bifurcation analysis using the AUTO. (Note that results of bifurcation analysis have already

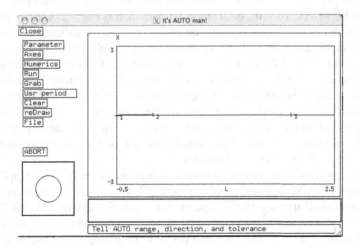

Fig. 1.21 The AUTO Window. Equilibrium points are drawn as a function of the parameter L. *Thick and thin lines* show stable and unstable equilibrium points, respectively. The labels 1 and 3 are the starting and ending points, respectively. The label 2 denotes the Hopf bifurcation point particularly

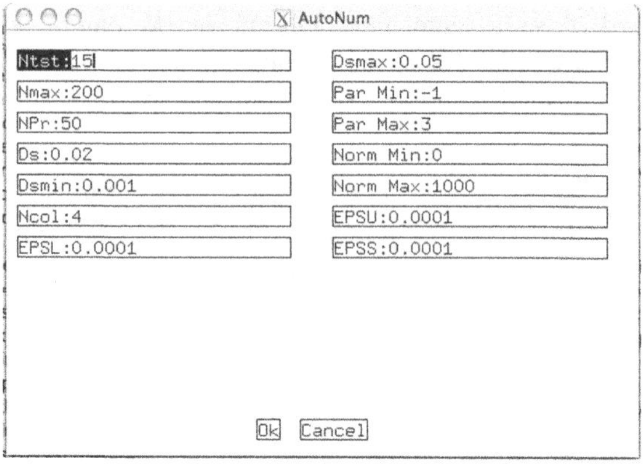

Fig. 1.22 AUTO Numerics Window. We can change the parameter values that AUTO uses in the numerical bifurcation analysis

been drawn, which will not appear at start-up.) By Axes and hI-lo in the left column of the AUTO window, we can set the variable to be plotted, the main bifurcation parameter (L here) (and the second parameter also) as well as the maximum and minimum scales of periodic orbits. If we click on Numerics, a new window such as Fig. 1.22 will appear. By this window, we set several parameters for numerical bifurcation analysis. For example, set the value of Dsmax to 0.05 which is the maximum step size of changing the bifurcation parameter. Also set the values of Par Min and Par Max to −1 and 3, respectively. These values denote the range where the main bifurcation parameter (L) should be changed. Then press on OK to close the window.

Now we are ready to obtain a bifurcation diagram. To do so, just click on Run and Steady state. Then the bifurcation diagram such as Fig. 1.21 can be obtained. The line with labels 1, 2 and 3 denote equilibrium points. The differential equations described in the hopf.ode file (or (1.49)) have an equilibrium point at the origin $(0, 0)$ for *all values* of L (λ). Thus we can see that the X value of the line is zero irrespective of the L value on the abscissa. The labels 1 and 3 denote the starting and ending points of bifurcation analysis, respectively. The point labeled 2 is a special (bifurcation) point detected automatically. The thick and thin lines denote stable and unstable equilibrium points, respectively. We can see that the stability change at the (Hopf bifurcation) point 2 from which a branch of periodic orbits should be bifurcated (see (1.49) and its explanation).

Next we proceed to calculate the branch of periodic orbits from the Hopf bifurcation point 2. To do so, let us "grab" the point 2. If we click on Grab button, then a cross (cursor) will appear on the one-parameter bifurcation diagram. We hit Tab (or arrow) keys several times on a keyboard to move the cursor on the Hopf bifurcation (HB) point labeled by 2. Then we grab the point by hitting the enter or return key. Finally, click on Run and Periodic buttons to obtain the bifurcation diagram shown in Fig. 1.23. The closed circles (dots) denote the maximum and minimum

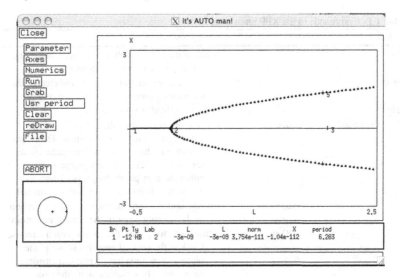

Fig. 1.23 AUTO Window. The bifurcation diagram of both equilibrium points and periodic orbits is shown. The *closed circles* (*dots*) bifurcated from the Hopf bifurcation point labeled by 2 show stable periodic orbits

values of stable periodic orbits. We can see that the Hopf bifurcation labeled by 2 is super-critical and stable periodic orbits are bifurcated through the Hopf bifurcation.

Note that, when we make an excursion along the curve of solutions (equilibrium points or periodic orbits) by clicking on Grab and by hitting tab and arrow keys, we can see several information (stabilities, parameter values, etc.) of the solution both in the lower-left corner and below the bifurcation diagram of the AUTO window.

Exercise 1.6. Create an ode file for (1.51). For the parameter values such as $L = -0.5$, $L = 0$ and $L = +0.5$, draw several obits with different initial values and vector fields to obtain similar X–Y phase planes to Fig. 1.12a–c. Also, obtain a bifurcation diagram of (1.51) similar to Fig. 1.13.   ∎

For a reference, Tables 1.1 and 1.2 show a partial list of XPP commands in the XPP main window and in the AUTO window, respectively. For the installation and the full list of XPP commands, visit

http://www.math.pitt.edu/~bard/xpp/xpp.html

Table 1.1 Commands in the XPP main window

Button		Action and use
ICs		A list of initial data will appear, and change the initial data you want to change
Param		A list of parameters will appear, and change the numerics parameters you want to change
quit		Quit the XPPAUT window
(I)nitialconds	(G)o	Use the initial data and the current numerics parameters to solve the equation. The output is drawn in the window and the data are saved for later use. The solution continues until either the user aborts by pressing Esc, the integration is complete, or storage runs out
	(L)ast	Use the end result of the most recent integration as the starting point of the current integration
	(M)ouse	Allows you to specify the values with the mouse. Click at the desired spot in a phase-plane picture. You must have a two-dimensional view and only variables along the axes
	m(I)ce	Allows you to choose multiple points with the mouse. Click Esc when done
(D)ir.field/flow	(D)irection fields	Choosing the direction field option will prompt you for a grid size. The two-dimensional plane is broken into a grid of the size specified and lines are drawn at each point specifying the direction of the flow at that point
(E)rase		Erasing the screen
(S)ing pts	(G)o	A new window about eigenvalue and its stability will appear
(V)iewaxes	(2)D	Set the variables to place on the axes, upper and lower limits of the two axes and the labels for the axes
(F)ile	(A)uto	Bring up the AUTO window if you have installed it
	(Q)uit	Quit the XPPAUT window

Table 1.2 Commands in the XPP AUTO window

Button and parameter		Action and use
`Close`		Close the AUTO window
`(P)arameter`		A list of parameters will appear. Type in the names of the parameters you want to use
`(A)xes`	`h(I)-lo`	Plot both the max and min of the chosen variable (convenient for periodic orbits)
	`(T)wo par`	Plot the second parameter versus the primary parameter for two-parameter continuations
`(N)umerics`	`Ntst`	This is the number of mesh intervals for discretization of periodic orbits. If you are getting bad results or not converging, it helps to increase this
	`Nmax`	The maximum number of steps taken along any branch. If you max out, make this bigger
	`Ds`	This is the initial step size for the bifurcation calculation. The sign of Ds tells AUTO the direction to change the parameter
	`Dsmin`	The minimum step size (positive)
	`Dsmax`	The maximum step size. If this is too big, AUTO will sometimes miss important points so if it seems to miss a stability transition, or if the diagram is jagged, decrease this
	`Par Min`	This is the left-hand limit of the diagram for the principle parameter. The calculation will stop if the parameter is less than this
	`Par Max`	This is the right-hand limit of the diagram for the principle parameter. The calculation will stop if the parameter is greater than this
`(R)un`	`(S)teady state`	Starting at a new steady state
	`(P)eriodic`	Compute the branch of periodics emanating from the Hopf point
	`(T)wo Param`	Compute a two parameter diagram of the selected points
`(G)rab`		Select a special point
`(U)sr period`	`Number (0-9)`	Choose the number of points you want to fix, and the fixed points will be labeled
`(C)lear`		Clear the diagram
`re(D)raw`		Redraw the diagram
`(F)ile`	`(P)ostscript`	Make a hard copy of the bifurcation diagram

Chapter 2
The Hodgkin–Huxley Theory of Neuronal Excitation

Hodgkin and Huxley (1952) proposed the famous Hodgkin–Huxley (hereinafter referred to as HH) equations which quantitatively describe the generation of action potential of squid giant axon, although there are still arguments against it (Connor et al. 1977; Strassberg and DeFelice 1993; Rush and Rinzel 1995; Clay 1998). The HH equations are important not only in that it is one of the most successful mathematical model in quantitatively describing biological phenomena but also in that the method (the *HH formalism* or the *HH theory*) used in deriving the model of a squid is *directly applicable* to many kinds of neurons and other excitable cells. The equations derived following this HH formalism are called the *HH-type equations*.

The dynamical system theory is very useful to analyze and understand the dynamics of the HH equations. On the other hand, the HH equations have rich mathematical structures and give many insights to mathematics also. For example, there are still many studies and mathematical findings on the "classical" HH equations (Plant 1976; Rinzel 1978; Hassard 1978; Troy 1978; Rinzel and Miller 1980; Labouriau 1985, 1989; Hassard and Shiau 1989; Shiau and Hassard 1991; Bedrov et al. 1992; Guckenheimer and Labouriau 1993; Labouriau and Ruas 1996; Fukai et al. 2000a,b). This chapter gives an overview of the dynamics of the HH equations and of the mechanism of the action potential generation.

2.1 What is a Neuron? Neuron is a Signal Converter

Figure 2.1 illustrates a shape and a function of neurons schematically. Our brain is a complicated network of a tremendous number of *neurons*. Right figure shows a *small* network of three neurons. A neuron has a very special shape which is much different from usual sphere-shaped or disklike cells. A *soma* is the main body of neuron from which a long cable called an *axon* is extended. Neurons transmit and exchange electric signals called *action potentials* or *spikes*, each other. (The generation of a spike is also called as the *excitation* or the *firing* of a neuron.) Neurons receive the spikes at a *synapse* which is a connection between neurons. Then, the electric signals or information is transmitted in the direction from a *dendrite* to an axon. The upper-left panel of Fig. 2.1 illustrates the waveform of action potentials. Action potential or a spike has an amplitude of about 100 mV.

S. Doi et al., *Computational Electrophysiology*,
DOI 10.1007/978-4-431-53862-2_2, © Springer 2010

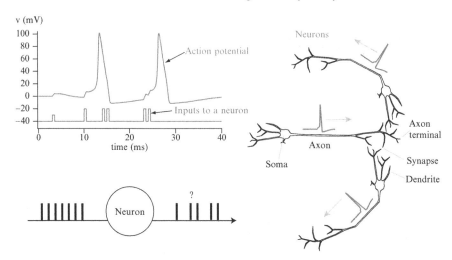

Fig. 2.1 Diagrams illustrating: A network of three neurons which exchange electric signals called action potentials, each other (*right*). Waveform of action potentials (*upper left*). Neuron as a device which converts input signals to output signals (*lower left*)

Typical neurons do not generate any spikes without input signals (i.e. spikes from other neurons). A sufficiently large input pulse causes a neuron to generate an output spike, as illustrated in the upper-left panel, whereas no output spike is generated by a small input (the first pulse in the lower trace of the upper-left panel). Therefore, a neuron possesses a *threshold* or *all-or-none* characteristic. There is a special period or timing called the *refractory period* (the timing of the downstroke of the action potential) in which the neuron cannot produce any output spike even though sufficient amount of inputs (the third and fourth pulses in the panel) were put in the neuron. Thus, we can consider a neuron as a device which transforms or converts the train of input spikes to a train of output spikes (see the lower-left panel of Fig. 2.1).

2.2 The Hodgkin–Huxley Formulation of Excitable Cell Membranes

This section briefly explains the framework of the HH formalism to model the action potential generation of neurons and of other excitable cells.

Biological cells, including neurons, are enclosed by a *plasma membrane* or simply *membrane* which separates the intracellular and extracellular water-containing media. The cell membrane consists of lipid bilayer, as shown in Fig. 2.2. There are various ions in both the intra- and extra-cellular regions. The concentrations of ions, however, are much different between the intra- and extra-cellular regions. For example, the concentration of K^+ ion is high and low in the intra- and extra-cellular regions, respectively. On the contrary, that of Na^+ ion is low and high

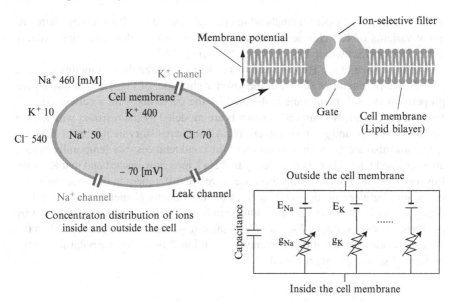

Fig. 2.2 The equivalent-circuit formulation of a cell membrane and ionic channels by Hodgkin and Huxley

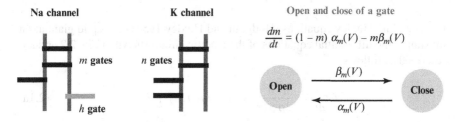

Fig. 2.3 Diagrams explaining the gate dynamics

in the intra- and extra-cellular regions, respectively. Usually, the resistance of the membrane is very high and the membrane acts as an insulator to the movement of ions. If the electrical potential at the inside surface of the cell membrane is compared to the potential at the outside surface, there is a potential difference or voltage called the *transmembrane potential* or simply the *membrane potential.*

In the membrane, there are holes through which ions can move in and out. Such a hole is called an *ion channel* and consists of *membrane proteins.* Ion channels are not simple holes (pores) or passive resistors through which ion flux flows. Ion channels are selective for a particular ion. For example, an ionic channel named Na^+ channel can pass only Na^+ ions. Also, ion channels are dynamic and sensitive to the membrane potential and to other factors. Namely, they open and close depending on such factors. An ion channel has several *gates* and the opening and closing of the ion channel are controlled by the gates as shown in Fig. 2.3. Note that, there are various types of ionic channels which pass a specific ion. Particularly, there are many variants of K^+ channels classified by their various characteristics (Adams 1982; Crill and Schwindt 1983; Llinas 1988; Hille 1992).

The basic idea of the HH formalism is to just recognize the cell membrane as a simple electric circuit as shown in the lower-right panel of Fig. 2.2. The capacitive property of the cell membrane is denoted by the *capacitor* with a certain *capacitance* in the circuit. Na^+ and K^+ channels are modeled by the *resistors* which have *conductances* g_{Na} and g_K, respectively. Note that the resistors are not linear but nonlinear, and also are dynamic: the values of the conductances vary temporally, which are explained later. There is a tendency that Na^+ ions flow inward and that K^+ ions flow outward the cell membrane because there are differences in the concentrations of the ions between inside and outside of the membrane. Namely, ions have a tendency to move down their concentration gradients. Such a tendency is denoted by the batteries with voltages E_{Na} and E_K in the circuit. These voltages depend on the inside–outside concentration difference in each ion. Notice that the polarities of the batteries E_{Na} and E_K are reversed.

2.3　Nonlinear Dynamical Analysis of the Original HH Equations

The Hodgkin–Huxley equations (Hodgkin and Huxley 1952) of a squid giant axon are simply the differential equations of the electric circuit shown in Fig. 2.2 and are described as follows:

$$C \frac{\partial v}{\partial t} = \frac{a}{2\rho} \frac{\partial^2 v}{\partial x^2} + G(v, m, n, h) + I_{\text{ext}}, \qquad (2.1\text{a})$$

$$\frac{\partial m}{\partial t} = \alpha_m(v)(1 - m) - \beta_m(v)m, \qquad (2.1\text{b})$$

$$\frac{\partial n}{\partial t} = \alpha_n(v)(1 - n) - \beta_n(v)n, \qquad (2.1\text{c})$$

$$\frac{\partial h}{\partial t} = \alpha_h(v)(1 - h) - \beta_h(v)h; \qquad (2.1\text{d})$$

$$\begin{aligned} G(v, m, n, h) &= I_{Na}(v, m, h) + I_K(v, n) + I_L(v) \\ &= \bar{g}_{Na} m^3 h (V_{Na} - v) + \bar{g}_K n^4 (V_K - v) + \bar{g}_L (V_L - v); \end{aligned} \qquad (2.2)$$

$C = 1\,\mu\mathrm{F\,cm^{-2}}, \; \bar{g}_{Na} = 120\,\mathrm{mS\,cm^{-2}}, \quad \bar{g}_K = 36\,\mathrm{mS\,cm^{-2}}, \quad \bar{g}_L = 0.3\,\mathrm{mS\,cm^{-2}},$
$V_{Na} = 115\,\mathrm{mV}, \quad V_K = -12\,\mathrm{mV}, \quad V_L = 10.599\,\mathrm{mV};$

$$\alpha_m(v) = \frac{0.1(25 - v)}{\exp[(25 - v)/10] - 1}, \quad \beta_m(v) = 4e^{-v/18},$$

$$\alpha_n(v) = \frac{0.01(10 - v)}{\exp[(10 - v)/10] - 1}, \quad \beta_n(v) = 0.125e^{-v/80},$$

$$\alpha_h(v) = 0.07\exp[-v/20], \quad \beta_h(v) = \frac{1}{\exp[(30 - v)/10] + 1};$$

where v (mV) is the membrane potential. Equation (2.1a) simply denotes the Kirchhoff's law. (In (2.1), a neuron is considered as a cylinder-shaped cell and the dependence of v on its position x is also taken into account. The term $(a/2\rho)(\partial^2 v/\partial x^2)$ denotes the diffusions of ions along the axis of the cylinder. However, either the shape of neurons or the x-dependence are not considered in this book.) I_{Na} and I_K are the currents through Na^+ and K^+ channels, respectively. The current I_L is the leak current and denotes all residue currents through a cell membrane other than Na^+ and K^+ currents.

As seen from (2.2), the Na^+ current I_{Na} is denoted by $\bar{g}_{Na}m^3h(V_{Na} - v)$ which takes a form of (*Conductance*) × (*Voltage*): Ohm's law. The voltage V_{Na} is called the *Nernst potential* or the *equilibrium potential* or sometimes the *resting potential* (do not confuse with the resting potential of whole membrane) of Na^+ ion. The Nernst potential is the potential where the tendency of ions to move down their concentration gradient is exactly balanced with the force by the electric potential difference; no Na^+ current flow through the Na^+ channel when $v = V_{Na}$. $\bar{g}_{Na}m^3h$ (= g_{Na} in the circuit of Fig. 2.2) denotes the conductance of Na^+ channel where the constant $\bar{g}_{Na}$ is called the *maximum conductance* of the channel and m^3h denotes dynamic or temporal change of the conductance. In the K^+ current I_K, the term n^4 denotes the temporal change of K^+ channel conductance.

The variables m, n and h take a (dimensionless) value between zero and unity, and are called the *gate variables*. As seen from the left panel of Fig. 2.3, it is assumed that Na^+ channel possesses three m-gates and single h-gate whereas K^+ channel four n-gates. In the HH formalism, it is also assumed that the variables m, n and h denote the probabilities that corresponding gates are open. The dynamic opening and closing of gates obey a simple (linear) process described by (2.1b–d) where $\alpha_m(v)$ ($\beta_m(v)$) is the rate "constant" for changing from closed (opened) state to opened (closed, resp.) state, as shown in the right panel of Fig. 2.3. Note that the rate "constant" $\alpha_m(v)$ and $\beta_m(v)$ are not actually the constants but the functions which depend on the membrane potential v. m and h are also called the activation and inactivation variables of Na^+ ionic channel, respectively, while n the activation variable of K^+ channel. The reason of this naming is because m and n are the increasing functions of v while h the decreasing one. I_{ext} ($\mu A\,cm^{-2}$) is the constant current externally applied to a neuron. The constants a and ρ are the radius and resistivity of the "cylindrical" axon, respectively. Throughout this book, we do not treat either such a cylindrical axon or the partial differential equation (2.1) which is an *infinite-dimensional* dynamical system. For the rich and complicated dynamics of such neuronal cable equations, see Carpenter (1977), Horikawa (1994), Kepler and Marder (1993), Rinzel and Keener (1983), Poznanski (1998), and Yanagida (1985, 1987, 1989).

2.3.1 Simple Dynamics of Gating Variables

In the following, we assume that the membrane potential v is spatially constant and omit the spatial derivative in (2.1); we consider the following HH equations for a *space-clamped* squid giant axon:

$$C\frac{dv}{dt} = G(v, m, n, h) + I_{\text{ext}}, \tag{2.3a}$$

$$\frac{dm}{dt} = \frac{1}{\tau_m(v)}(m^\infty(v) - m), \tag{2.3b}$$

$$\frac{dn}{dt} = \frac{1}{\tau_n(v)}(n^\infty(v) - n), \tag{2.3c}$$

$$\frac{dh}{dt} = \frac{1}{\tau_h(v)}(h^\infty(v) - h), \tag{2.3d}$$

where

$$\tau_x(v) = \frac{1}{\alpha_x(v) + \beta_x(v)}, \quad x^\infty(v) = \frac{\alpha_x(v)}{\alpha_x(v) + \beta_x(v)}, \quad x = m, n, h. \tag{2.4}$$

(The word "space-clamped" means that both the shape of a neuron and the dependence of the membrane potential v on the spacial position x are ignored while both were taken into account in (2.1).)

Figure 2.4 shows an example of a numerically solved solution of the HH equations (2.3). Panel (a) is a waveform of the membrane potential. A pulsatile input applied at a time $t = 5$ ms induces an action potential. Panel (b) is the waveforms of the gating variables m, n and h. Total membrane current, Na current, K current and leak current are shown in (c)–(f), respectively.

The HH equations (2.3) are nonlinear differential equations with four variables and apparently look very complicated. Equations (2.3b–d) which describe the dynamics of gating variables, however, share a simple common structure. Functions $\tau_x(v)$ and $x^\infty(v)$, $x = m, n, h$ depend on the membrane potential v and thus vary temporally with the temporal change of v. If we assume that the functions do not depend on v ($\tau_x(v) \equiv \tau_x$, $x^\infty(v) \equiv x^\infty$), then (2.3b–d) reduce to a linear differential equation

$$\frac{dx}{dt} = \frac{1}{\tau_x}(x^\infty - x), \quad x = m, n, h,$$

whose analytical solution with an initial value $x = x_0$ is

$$x(t) = \exp(-t/\tau_x)(x_0 - x^\infty) + x^\infty.$$

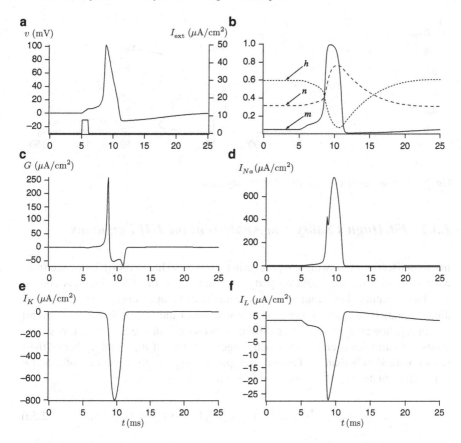

Fig. 2.4 Solution of the Hodgkin–Huxley equations

The variable $x(t)$ approaches x^∞ at a speed depending on a time constant τ_x. In (2.3b–d), although they are function of v and we cannot solve the equation analytically, $\tau_x(v)$ still preserves a role of "time constant" and $x^\infty(v)$ is the steady-state function to which the variable x asymptotically approaches in a steady (or stationary) state.

Figure 2.5a shows the functions $x^\infty(v)$, $x = m, n, h$. In spite of the complicated functional forms, these functions are all monotonic functions and have the shape of so-called "sigmoid function." $m^\infty(v)$ and $n^\infty(v)$ are increasing functions and thus variables m and n are activation variables while $h^\infty(v)$ is a decreasing function and h is an inactivation variable. Figure 2.5b shows the functions $\tau_x(v)$, $x = m, n, h$ which change depending on v. The function $\tau_m(v)$, however, much smaller than $\tau_n(v)$, $\tau_h(v)$ in the whole range of v. In the following, let us explore the dynamics of the HH equations regarding this time-scale difference.

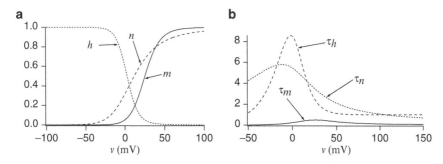

Fig. 2.5 Various functions in the Hodgkin–Huxley equations

2.3.2 FitzHugh's Subsystem Analysis of the HH Equations

In this subsection, following the pioneering paper FitzHugh (1960), let us see from
what dynamics the threshold property of a neuron or the HH equations comes.
The HH equations have four variables and it is difficult to observe the full (four-
dimensional) state space directly. The separation of the full state space to several
subspaces, however, resolves this difficulty. As shown in Fig. 2.5, the "times constants" of variables n and h are much bigger than that of m; n and h change their
values more slowly than m. Thus we (temporarily) ignore the dynamics of n and h
in the HH equations and consider the following system:

$$C\frac{dv}{dt} = \bar{g}_{Na}m^3h(V_{Na} - v) + \bar{g}_K n^4(V_K - v) + \bar{g}_L(V_L - v), \qquad (2.5a)$$

$$\frac{dm}{dt} = \frac{1}{\tau_m(v)}(m^\infty(v) - m). \qquad (2.5b)$$

(We call this system as the v–m subsystem.) In the v–m subsystem, v and m are the
dynamic variables while h and n are set in suitable values as a "parameter."

By using the v–m subsystem, let us explain the firing process in the HH equations
in the following order:

> quiescent state → depolarization → decrease of h → increase of n
> → repolarization .

Figure 2.6a shows the v–m phase plane of (2.5) when the values of h and n are fixed
to that of the quiescent state (a stable equilibrium point) of the HH equations (2.3).
The m-nullcline (a curve in which $dm/dt = 0$: $m = m^\infty(v)$) is a sigmoidal mono-
tonically increasing function. (For more explanation of nullclines, see Chap. 3.) We
can see that the v-nullcline (dotted curve) and m-nullcline (solid curve) intersect in
the three points v_1^*, v_2^* and v_3^* which are equilibrium points of the v–m subsystem
(panel b is the magnification of lower-left region of a). The leftmost equilibrium

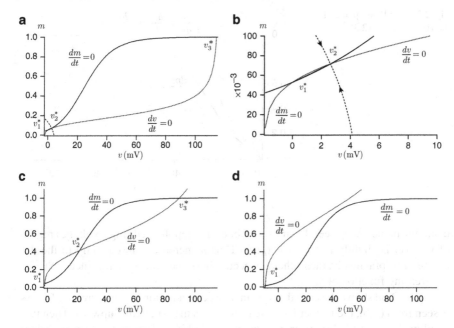

Fig. 2.6 Phase plane of the v–m subsystem (2.5) of the HH equations. (**a**) $n = 0.317677$, $h = 0.596120$. (**b**) Magnification of (**a**). (**c**) $n = 0.317677$, $h = 0.02$. (**d**) $n = 0.5$, $h = 0.02$

point v_1^* corresponds to the quiescent state (a stable equilibrium point) and the middle point v_2^* is a saddle point whose stable manifold (the broken curve tending to v_2^*) forms a threshold between exciting and non-exciting, which means that the HH model when h and n are fixed to the values of quiescent state is the type-I neuronal model with a strict threshold. If a sufficiently large stimulus is applied to a neuron in a quiescent state v_1^*, the state point moves rightwards beyond the stable manifold of v_2^* and then goes towards the rightmost equilibrium v_3^* which corresponds to the depolarized state of a neuron. If a membrane potential v increases, h is decreased (after a slight delay) because the variable h tends to the function $h^\infty(v)$ which is a decreasing function of v.

Figure 2.6c is the phase plane with a smaller value of h. In the v–m subsystem (2.5), the v-nullcline does depend on both h and n while the m-nullcline does not depend on them. From (2.5a), the v-nullcline is obtained by

$$m^3 = \frac{\bar{g}_K n^4 (v - V_K) + \bar{g}_L (v - V_L)}{\bar{g}_{Na} h (V_{Na} - v)} \tag{2.6}$$

from which we can see that the decrease of h moves the v-nullcline upward (the shape of v-nullcline is also changed). Note that we consider this equation in the range of $0 \leq m \leq 1$, thus the right-hand side is positive. As a result of displacement of the v-nullcline, three intersections of the v-, m-nullclines become more clearly

Fig. 2.7 v–m Phase plane of
the HH equations (2.3)

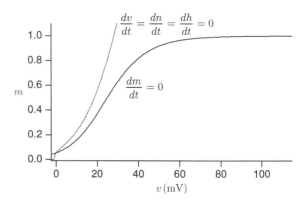

distinguishable. As is seen from Fig. 2.5b, in the depolarized (high-voltage) region
of v, $\tau_n(v)$ is slightly larger than $\tau_h(v)$. Thus n increases subsequently to the decrease of h (practically, these changes occur almost simultaneously) since $n^\infty(v)$ is
a increasing function of v.

Figure 2.6d shows the case that n is increased in addition to the decrease of h. As
is seen from (2.6), the further increase of n moves the v-nullcline upward. Then two
equilibria v_2^* and v_3^* disappear by a saddle-node bifurcation and only equilibrium
v_1^* remains. As a result of disappearance of those equilibria, the state point near the
equilibrium v_3^* (depolarized state) cannot stay there and then changes its direction
toward the equilibrium v_1^* (quiescent state). After this process, h and n change to
increase and decrease, respectively, and then return to the state of Fig. 2.6a.

Figure 2.7 shows the v–m phase plane of the *full* HH equations (2.3) (projec-
tion of the four-dimensional state space to v–m phase plane). The sigmoid-like
curve is the m-nullcline and is same as the case of v–m subsystem. The
other curve denotes the intersection of v, n, h-nullclines; a curve such that
$dv/dt = dn/dt = dh/dt = 0$:

$$m^3 = \frac{\bar{g}_K\{n^\infty(v)\}^4(v - V_K) + \bar{g}_L(v - V_L)}{\bar{g}_{Na}h^\infty(v)(V_{Na} - v)}.$$

The intersection of these two nullclines corresponds to the equilibrium point of the
original HH equations (2.3) and thus the HH equations have a unique equilibrium.
This implies that in a strict sense the original HH equations are a type-II neuronal
model without a distinct threshold. As is seen above, however, in a very short time
range (i.e. if we ignore the temporal change of h and n) the HH equations behave
like a type-I neuronal model which has a distinct threshold. Thus we can understand
why the HH equations have a relatively sharp threshold although they do not have
any threshold in a strict sense.

2.3.3 Dynamic I–V Relation of the Squid Giant Axon Membrane

In this subsection, let us investigate the HH equations from more global viewpoint without entering into its detailed dynamics. Figure 2.8 shows dynamic or transient current–voltage relation of the HH equations; total membrane currents G when a time t elapses after the membrane voltage is instantaneously changed to v (mV) from the quiescent state are plotted for various v values. These current–voltage relations are shown for various values of the time t. At a time $t = 0.02$ ms, the relation is almost linear. After some time elapses, negative-resistor characteristics appear and the current–voltage relation becomes N-shaped. After sufficiently long time elapses, the relation shows a rectifier characteristics in which only outward current can be flowed.

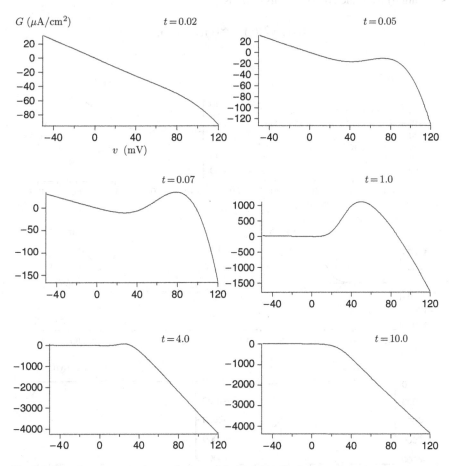

Fig. 2.8 Transient current–voltage relations of the Hodgkin–Huxley membrane

In order to see how these current–voltage relations are generated, let us decompose the total membrane current to its components: I_{Na} and I_K where the (small) leak current is ignored. Figure 2.9 is the similar voltage–current relations to Fig. 2.8 for Na and K channels. When $t = 0.02$, the relation of K channel (dotted curve) is almost linear. After time elapses, the K current in a low-voltage range becomes small and finally, the current–voltage relation becomes a rectifier. In the case of Na channel (solid curve), when $t = 0.02$, it does not flow much current (almost flat). After suitable time elapses, it shows large nonlinear characteristics and then finally becomes almost flat again. The combination of these two current–voltage relations makes the I–V relation of the total current in Fig. 2.8.

The above current–voltage relations are obtained by numerically solving the HH equations (2.3). The relation when $t = 10.0$ is considered as a steady-state current–voltage relation. This relation can be obtained directly (without numerical simulation) from the HH equations (2.3) as

$$G(v, m^\infty(v), n^\infty(v), h^\infty(v)).$$

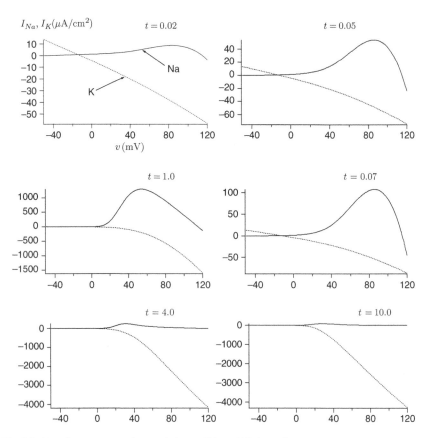

Fig. 2.9 Transient current–voltage relations of Na and K channels

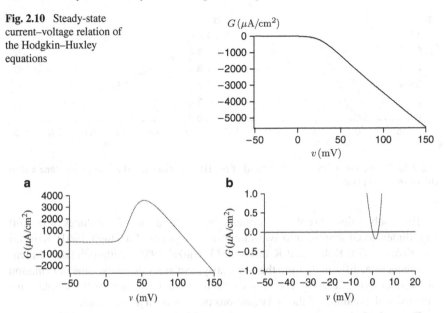

Fig. 2.10 Steady-state current–voltage relation of the Hodgkin–Huxley equations

Fig. 2.11 (a) Steady-state current–voltage relation of the v, m-subsystem of the HH equations. (b) Magnification of (a)

This steady-state current is plotted as a function of v in Fig. 2.10. Similarly to the $t = 10$ case of Fig. 2.8, we can see the rectifier characteristics.

Figure 2.11 shows the steady-state current of the v–m subsystem (2.5):

$$G(v, m^\infty(v), n^*, h^*)$$

as a function of v where n^* and h^* denote the quiescent-state values of n and h, respectively. Panel (b) is the magnification of (a), from which the current–voltage curve intersect with the horizontal line at three points which correspond to the three equilibrium points of the phase plane of Fig. 2.6a. From this observation, we can also see the threshold property of the HH equations.

Comparing Fig. 2.11 to the transient current–voltage relation of Fig. 2.8, we confirm that the v–m subsystem approximates the dynamics of the full HH equations (2.3) in the time scale of $t \leq 1$ ms.

2.3.4 Dimension Reduction of the HH Equations

In the v–m subsystem analysis, we investigated the v–m phase plane of the HH equations by fixing the values of n and h. Namely, we explored the dynamics of HH equations by decomposing the four-dimensional full phase space into several n, h-fixed slices.

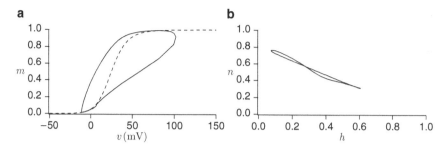

Fig. 2.12 Projections of the orbit (solution) of the HH equations to (**a**) a v–m phase plane and to (**b**) a h–n phase plane

This subsection briefly describes the method which reduces the full four-dimensional system into two-dimensional system (FitzHugh 1961; Krinskii and Kokoz 1973; Kokoz and Krinskii 1973; Rinzel 1985). Although such dimension reduction, differently from the v–m subsystem analysis, loses some information on original dynamics, it is useful to catch the essential feature of the whole four-dimensional dynamics of the HH equations on a reduced phase plane.

Figure 2.12 shows the projections of an orbit (solution) $(v(t), m(t), h(t), n(t))$ of the HH equations to (a) a v–m phase plane and to (b) a h–n phase plane. From panel (a), we can see that the orbit in the region of $m \approx 0$ or $m \approx 1$ moves close to the m-nullcline $m = m^\infty(v)$ (broken curve). Panel (b) shows that the orbit moves restricted in a certain line on the h–n phase plane.

From these observations, we reduce the HH equations (2.3) in two steps:

1. Suppose that the variable m which follows (2.3b) is settled in its steady-state value: $m = m^\infty(v)$ since m is the fast-changing variable (τ_m is small). Thus we ignore (2.3b) and substitute $m^\infty(v)$ for m in (2.3a).
2. Approximate the orbit on the h–n phase plane by a line $n = 0.8(1 - h)$; we consider that the variable n linearly depends on h. Thus we ignore (2.3c) and substitute $0.8(1 - h)$ for n in (2.3a).

From these reduction steps, we obtain the reduced equations:

$$C\frac{dv}{dt} = G(v, m^\infty(v), 0.8(1 - h), h) + I_{\text{ext}}, \tag{2.7a}$$

$$\frac{dh}{dt} = \alpha_h(v)(1 - h) - \beta_h(v)h. \tag{2.7b}$$

This model has only two dynamic variables v and h, and thus has an advantage that we can analyze the neuronal dynamics on a phase plane rather than the four-dimensional phase space of the original HH equations.

Figure 2.13a shows an example of the membrane potential waveforms of the original HH equations (2.3) (dotted curve) and of the reduced model (2.7) (solid curve). The upstroke of the membrane potential of the reduced model is slightly faster than that of the HH equations and the peak value is also bigger than the HH equations.

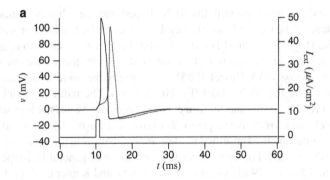

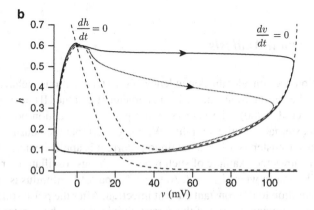

Fig. 2.13 Comparison of (**a**) membrane potential waveforms and (**b**) solution orbits in the phase plane, between the dimension-reduced model (*solid curve*) and the original HH model (*dotted curve*)

This is because we substituted $m^\infty(v)$ for m and thus the activation process of Na channel has been slightly accelerated. Except for these small differences, the two waveforms of the original and the reduced model are quite similar. This similarity is surprising since the dimensions of the original and the reduced model are much different.

Figure 2.13b shows the solution orbit of both models on the v–h phase plane. The solid curve denotes the orbit of the reduced model (2.7), and the dotted curve the original HH equations (2.3). The v-nullcline

$$G(v, m^\infty(v), 0.8(1-h), h) = \bar{g}_{Na}(m^\infty(v))^3 h(V_{Na} - v)$$
$$+ \bar{g}_K(0.8(1-h))^4(V_K - v) + \bar{g}_L(V_L - v) = 0$$

and the h-nullcline

$$h = h^\infty(v) = \frac{\alpha_h(v)}{\alpha_h(v) + \beta_h(v)}$$

are also shown by broken curves. We can see the similarity of orbits of the two models except for the slight difference in the action potential upstroke.

Topological features (one nullcline is N-shaped and the other is monotonic) of the phase plane of the reduced model resemble to that of the Bonhoeffer–van der Pol (BVP) or FitzHugh–Nagumo (FHN) model (FitzHugh 1961; Nagumo et al. 1962). In fact, the assumptions (i) and (ii) which are used to derive the reduced model (2.7) (Krinskii and Kokoz 1973; Rinzel 1985) is essentially the same as the one used by FitzHugh to derive the BVP model (FitzHugh 1961). The reduced model (2.7) by Krinskii and Kokoz (1973) and by Rinzel (1985) is derived by some logical process while the BVP model is derived a priori. Thus, the relation of physiological parameters of the original and the simplified models are much clear in the model (2.7) rather than in the BVP model. For more systematic reduction of general HH-type models, see Abbott and Kepler (1990), Golomb et al. (1993), and Kepler et al. (1992).

2.3.5 Bifurcation Analysis

In this subsection, we consider the dependence of the HH equations' behavior on the parameter I_{ext}; the constant-current-transfer characteristic of the HH neuron (Rinzel 1978; Guttman et al. 1980). The neuron model produces an action potential with response to an external pulsatile stimulus. We can expect that the neuron generates action potentials persistently when a continuous current is applied externally.

Figure 2.14 shows the example of such repetitive firings (oscillation or periodic orbit) of the HH neuron when $I_{ext} = 7$. At $t = 62$, a pulsatile stimulus is applied to the neuron in addition to the constant current injection. After the pulse stimulus, the repetitive firing is stopped in spite of the persistent injection of the constant current. This means that an repetitive firing (stable limit cycle) and a quiescent state (stable equilibrium) coexist when $I_{ext} = 7$. The state point of the HH neuron is moved from one attractor (limit cycle) to an another one (equilibrium) by the external pulse.

We note that the applied pulse is depolarizing current; not only inhibitory input but also excitatory input can inhibit such repetitive firing since it is a nonlinear oscillation. The timing (phase) when the pulse is applied is important.

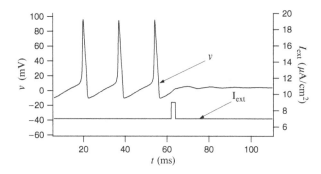

Fig. 2.14 Example of a membrane potential waveform of the HH equations when an external constant current is applied: $I_{ext} = 7 \mu A\ cm^{-2}$. A pulse is applied to the model in addition to the constant current at $t = 62$

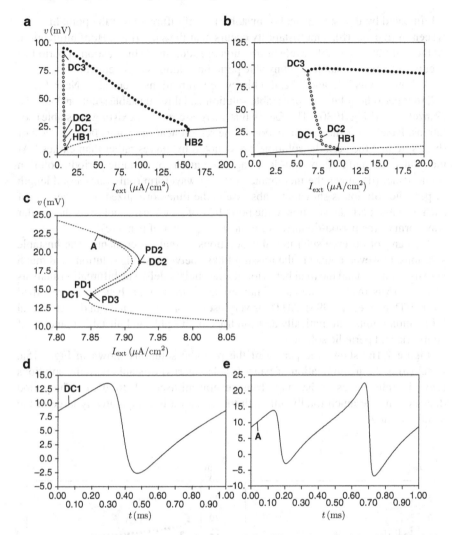

Fig. 2.15 (**a**) One-parameter bifurcation diagram of the HH equations. (**b**) Magnification of (a). (**c**) Magnification of (b). (**d, e**) Membrane potential waveforms at the points DC1 and A of (c)

Figure 2.15a shows the dependence of the solution of the HH equations on the parameter I_{ext}; the v values of the stationary solution of the HH equations are plotted for various values of I_{ext} where the maximum value of v is plotted for a periodic (oscillatory) solution. Solid and dotted curves denote stable and unstable equilibria, respectively. The filled (open) circles denote stable (unstable, resp.) periodic solutions.

Panel (b) is the magnification of left part of (a) and we can verify the multistability of an equilibrium and a periodic solution when $I_{ext} = 7$ (ref. Fig. 2.14). At the point HB2 of panel (a), a stable periodic solution bifurcates from a equilibrium point by the (super-critical or stable) Hopf bifurcation. An unstable periodic solution is bifurcated by the (sub-critical or unstable) Hopf bifurcation at the point HB1. As is seen from panel (b), a multistability occurs near the sub-critical Hopf bifurcation. At the point DC3, a double-cycle bifurcation or saddle-node bifurcation of periodics occurs and a pair of stable and unstable periodic solutions is generated.

At both points DC1 and DC2, double-cycle bifurcations occur also. Near $I_{\text{ext}} = 7.9$, four periodic solutions (one stable solution and three unstable solutions) coexist (Rinzel and Miller 1980). The fact is that there are more coexisting (unstable) solutions. Panel (c) is the magnification of the region near the points DC1 and DC2 of (b). (Note that the periodic solutions are denoted by curves rather than circles.) At the points PD1 and PD2, period-doubling bifurcations of unstable periodic solution occur. Panel (d) shows the membrane potential waveform (only one-period length of periodic solution is shown and abscissa is the time normalized by its period) of such unstable periodic solution at the point DC1 of panel (c). Panel (e) is the similar waveform of the period-doubled solution at the point A of panel (c).

We can not observe such unstable solutions in real experiments. The unstable solutions, however, connect "the missing links" between stable solutions and much help us to understand the total behavior of neuronal models. The bifurcation analysis of Fig. 2.15 is made by the use of numerical bifurcation analysis software called AUTO (Doedel et al. 1995). AUTO is very useful software which can detect several bifurcation points automatically and can trace both stable and unstable branches of equilibria and periodic solutions.

Figure 2.16a shows the period of the periodic solutions shown in Fig. 2.15a. Panel (b) is the magnification of panel (a). The period of stable periodic solutions (closed circle) varies in the range from several milliseconds to 20 ms. The period does not change much totally although the variation is comparatively large in the small I_{ext} range.

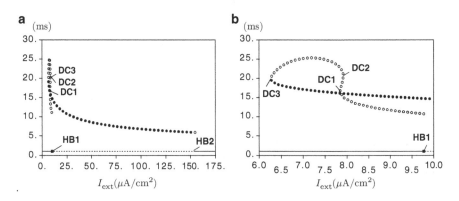

Fig. 2.16 (a) Period of the periodic solution shown in Fig. 2.15a and (b) Magnification of (a)

Chapter 3
Computational and Mathematical Models of Neurons

What are models? The HH equations (2.3) are often called a physiological model, whereas the models appeared in the following sections are simplified models or abstract models. However, there is no model in which any simplifications or abstractions have not been made. Of course, many features of real neurons are ignored even in the HH equations. All models have their applicability and limits to describe natural phenomena. Therefore, all types of models whatever simplified or physiological, have their own values to model real phenomena. Starting with the BVP or FHN model which is a simplification of the HH equations, this chapter proceeds to several neuronal models with higher abstractions which are useful to tract some essential features of neurons.

3.1 Phase Plane Dynamics in the Context of Spiking Neuron

Neurons generate action potentials or spikes as seen in the above chapter. In this section, let us consider the one of the simplest models of such spike generation. The Bonhoeffer–van der Pol (BVP) model is derived from the well-known van der Pol equation by FitzHugh (1961) which mimics well the qualitative behavior such as excitability, refractoriness, repetitive activity of the Hodgkin–Huxley model. The BVP model is also called as the FitzHugh–Nagumo (FHN) model because the model is equivalent to the electronic circuit with a tunnel diode proposed by Nagumo et al. (1962).

The (space-clamped) BVP model is expressed by the following set of differential equations:

$$\dot{x} = c(x - x^3/3 - y + I_{ext}), \tag{3.1a}$$

$$\dot{y} = (x - by + a)/c, \tag{3.1b}$$

where $\dot{x}$ means dx/dt. The variable x denotes the membrane potential of a neuron, y is a "recovery" variable which corresponds to the combination of Na^+ inactivation and K^+ activation of the HH model (2.3), and I_{ext} the current stimulus applied

S. Doi et al., *Computational Electrophysiology*,
DOI 10.1007/978-4-431-53862-2_3, © Springer 2010

externally. For the equations to mimic the behavior of a neuron, the parameters a, b and c are set in a certain range (FitzHugh 1961; Nagumo et al. 1962; Hadeler et al. 1976), and the conventional values are $a = 0.7$, $b = 0.8$, $c = 3.0$.

In the following, we explore the *neuronal* dynamics of the BVP equations using XPPAUT.

3.1.1 Excitability and Flows of the BVP Neuron Model

The following is the XPPAUT ode file of the BVP equations.

```
# bvp.ode
init x=1 y=0
x' = c*(x - x^3/3 - y + Iext)
y' = (x - b*y + a)/c
par Iext=0, b=0.8, c=3.0, a=0.7
@ xplot=t,yplot=x
@ total=100,dt=.03,xlo=0,xhi=10, ylo=-2,yhi=2
@ meth=runge-kutta
@ bound=200
@ autoxmin=0,autoxmax=1.8,autoymin=-3,autoymax=3
@ ntst=15 nmax=75 npr=200 parmin=-5 parmax=5 dsmax=0.1
   ds=0. 1 done
```

In this ode file, not only the differential equations but also the initial (default) values of several parameters for the execution of the numerical bifurcation analysis by AUTO are described.

Figure 3.1 shows the x-waveforms of solutions of the equations with various initial values: y value is fixed to zero, and x values are $x = \pm1$, ±0.5, ±0.4, ±0.3, ±0.2, ±0.1, 0. We can see that the solutions are mainly classified to two groups: Solutions with a positive initial value of x grow up and make a big wave at first, then they converge to a stable equilibrium point while other solutions decrease without such a big wave, then they also converge to the equilibrium. The variable x corresponds to the membrane potential of a neuron and the big wave corresponds to an action potential. Therefore, we can see that the solutions of the BVP equations are very sensitive to their initial values of x. The initial value of x corresponds to a current injection to a neuron; sufficiently large current injection to the BVP neuron model can produce an action potential, while small current injection cannot. These results demonstrate the excitability and the threshold characteristics of the BVP model. Strictly speaking, there is an intermediate solution between the solutions with and without action potential. This means that the threshold is not strict or that the response of the BVP neuron model is not *all-or-none*. We may call such a threshold a *quasi-threshold*.

Figure 3.1 can be obtained by using the XPPAUT as follows: Start the XPPAUT as shown in Fig. 1.16, and select or open the bvp.ode file. Then the XPPAUT main window will appear. Click on the ICs button which is the leftmost button on the top

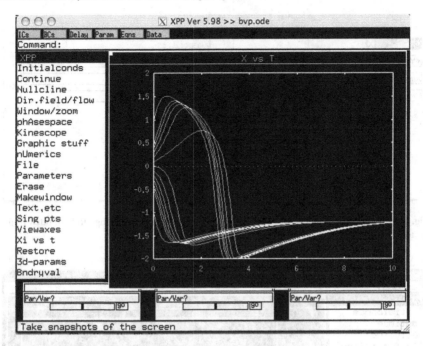

Fig. 3.1 Solutions of the BVP equations (3.1). $I_{ext} = 0$. Initial values are $y = 0$, $x = \pm 1, \pm 0.5$, $\pm 0.4, \pm 0.3, \pm 0.2, \pm 0.1, 0$

Fig. 3.2 Initial Data
window to set initial values
of the BVP equations

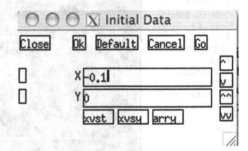

row of the XPP main window. Then the window as Fig. 3.2 will appear. If we set initial values and then click on (G)o, a waveform will be drawn. Iterations of this process may produce the figure.

As already explained in Sect. 1.5, in order to create a phase plane which corresponds to Fig. 3.1, click on the Viewaxes button and then 2D button, the 2D View window will comes up. If we change the axes and their ranges to such values as Fig. 3.3, then we can obtain the phase plane shown in Fig. 3.4.

To draw the vector field (or direction field) as Fig. 3.5, just click on the Dir.field/flow and (D)irect Field buttons. Then, in the second row of the XPP main window, a message such as "Grid:10" will appear. If you accept this default value, just hit an enter (or return) key, if not accept, modify the value as you want and hit enter.

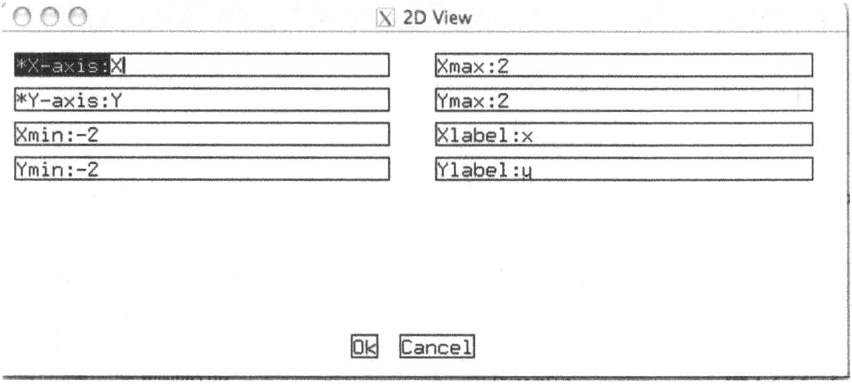

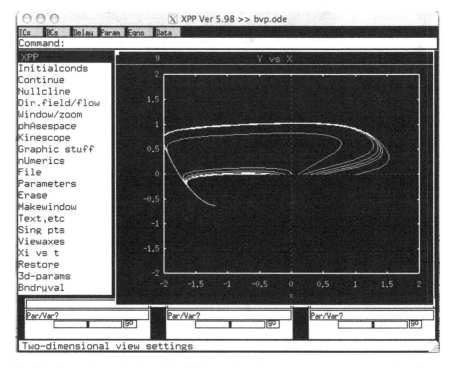

Fig. 3.3 The 2Dview window. We can change the variables and their range to be plotted in the X- and Y-axes

Fig. 3.4 Phase plane and orbits of the BVP equations. Initial values are $y = 0$, $x = \pm 1$, ± 0.5, ± 0.4, ± 0.3, ± 0.2, ± 0.1, 0

In Fig. 3.5, *nullclines* are also drawn. The x-nullcline is the curve such that $\dot{x} = 0$. As seen from (3.1), this curve is the cubic curve: $y = x - x^3/3 + I_{ext}$. The y-nullcline is the curve such that $\dot{y} = 0$ and thus the line $x - by + a = 0$. Since $\dot{x} = 0$ ($\dot{y} = 0$) on the x- (y-) nullcline, trajectories cross the nullcline verti-

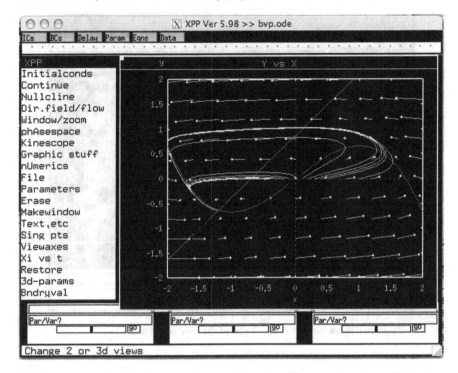

Fig. 3.5 Phase plane, vector field and nullclines of the BVP equations. Initial values of orbits are $y = 0$, $x = \pm 1, \pm 0.5, \pm 0.4, \pm 0.3, \pm 0.2, \pm 0.1, 0$

cally (horizontally, resp.). To draw the nullclines by XPP, just click on Nullcline and (N)ew. The intersection point of x- and y-nullclines is the equilibrium point. We can see that all trajectories approach the equilibrium point since this point is stable. We can verify this stability by the linearization and simple linear algebra (see Sect. 1.3).

Figure 3.5 shows the (quasi-) threshold characteristics of the BVP model again. We can see that the threshold exists near the "middle" branch of the cubic x-nullcline. The middle branch is, in a certain sense, unstable and thus it makes the sensitivity on initial values and the threshold characteristics, although we do not explain this in detail. Anyway, the geometric analysis of a phase plane (phase space) is very useful to analyze various neuronal characteristics of the BVP model.

3.1.2 Bifurcations in the BVP Neuron Model

Next, we proceed to the bifurcation analysis of the BVP equations. Let us restart XPPAUT and choose the bvp.ode file again, then an XPP main window will appear. If we click on the Initialconds button and then click on the (G)o button, a wave

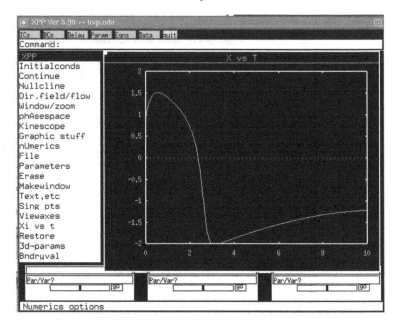

Fig. 3.6 XPP main window and the X vs. T plot of the BVP equations (3.1)

of X vs. T will be drawn as Fig. 3.6. After the integration (drawing of waveform) finished, we click on both Initialconds and (L)ast buttons (several times) to obtain enough convergence to a stable equilibrium point. We click on the File button and then click on the Auto button, an AUTO window will appear.

In the AUTO window, if we click on Run button and click on Steady state button, then a one-parameter bifurcation diagram will be drawn as Fig. 3.7. The abscissa is the bifurcation parameter I_{ext} and the ordinate is the x variable of the BVP equations (3.1). The x-coordinate of the equilibrium points (steady states) of the BVP equations are plotted as a function of the external stimulus current I_{ext}. Thick and thin curves shows stable and unstable equilibrium points, respectively. As I_{ext} increases, the stable equilibrium point becomes an unstable one at the point labeled by 2. This is the Hopf bifurcation point. If the stimulus current I_{ext} is further increased, the unstable equilibrium point recovers its stability at another Hopf bifurcation point labeled by 3.

In order to clarify what is happening between the two Hopf bifurcation points, we trace solutions which may be bifurcated through the Hopf bifurcation point 2. To do so, let us "grab" the point 2. If we click on Grab button, then a cross (cursor) will appear on the one-parameter bifurcation diagram. We use the Tab (or arrow) key on a keyboard to move the cursor on the Hopf bifurcation (HB) point labeled by 2. Then we grab the point by hitting the enter or return key. Finally, if we click on Run and Periodic buttons, the one-parameter bifurcation diagram on both steady states and periodic orbits shown in Fig. 3.8 will be drawn.

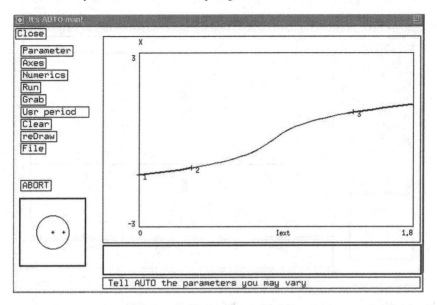

Fig. 3.7 One-parameter bifurcation diagram on steady states (equilibrium points) of the BVP equations (3.1)

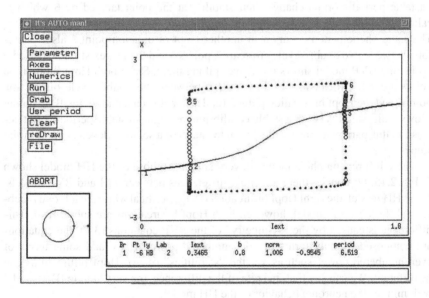

Fig. 3.8 One-parameter bifurcation diagram on *both* steady states and periodic orbits of the BVP equations (3.1)

The magnification of this bifurcation diagram near the left Hopf bifurcation point 2 is shown in Fig. 3.9. The open and closed circles denote unstable and stable periodic orbits, respectively. Unstable periodic orbits are bifurcated from

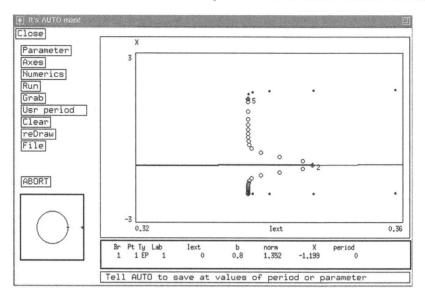

Fig. 3.9 Magnification of the one-parameter bifurcation diagram of Fig. 3.8

the Hopf bifurcation point 2: this Hopf bifurcation point is *sub-critical*. Those unstable periodic orbits change their stability at the point labeled by 5 which is called the *double-cycle bifurcation* or *saddle-node bifurcation of periodic orbits*. The same phenomenon occurs near another Hopf bifurcation point 3 also. Therefore, between two double-cycle bifurcation points 5 and 6, there exist stable periodic orbits: the BVP model shows repetitive spiking there. Since both Hopf bifurcation points are sub-critical, in the range of I_{ext} between the double cycle bifurcation point 5 and the Hopf bifurcation point 2 (and also between 3 and 6), the BVP model shows multi-stability: both a stable equilibrium point and a stable periodic orbit coexist. In this parameter range, the final fate (state) of a solution depends on its initial values.

These bifurcation phenomena are very similar to those of the HH model shown in Fig. 2.15. Of course, there are slight differences between HH and BVP models. In the HH model, the right Hopf bifurcation is super-critical while the left one is subcritical. In the BVP model, however, both Hopf bifurcations are sub-critical. This difference is caused by the "symmetry" of the BVP equations (3.1); the equations are composed by only terms of odd-numbered power. If we add some terms of even-numbered power such as x^2, the "stabilities" of two Hopf bifurcations can be different. We can say that, in spite of the difference, the very simple BVP model well mimics the neuronal behavior of the HH model.

In order to make the magnification shown in Fig. 3.9, click on Axes and hI-lo buttons, and change the scales in the window as shown in Fig. 3.10. Then, the clicking on Clear and reDraw buttons will give us Fig. 3.9.

Next, let us explore more detailed behavior of bifurcations of the BVP model (3.1). In particular, we perform the two-parameter bifurcation analysis.

Fig. 3.10 The AutoPlot
window for the scale change

Fig. 3.11 The Parameters
window for the change of
parameter values

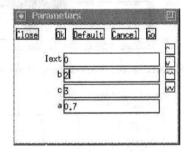

To do so, let us "restart" the XPP and choose the bvp.ode file again (Note that we do not actually need to restart the XPP. We can continue the analysis. But the restart is much simpler than the continuation). At first, in the XPP main window, we click on Param button and change the default value of b to 2, as shown in Fig. 3.11. If we click on Initialconds and (G)o buttons, and subsequently Initialconds and (L)ast buttons (a few times), then Fig. 3.12 will be drawn.

We click on File and Auto buttons to show a AUTO window. Then, if we click on Run and Steady state buttons, a one-parameter bifurcation diagram will be drawn as Fig. 3.13. The x-coordinates of equilibrium points are plotted as a function of I_{ext}. Differently from the previous bifurcation diagram of Fig. 3.7 where $b = 0.8$, the curve of equilibrium points is not monotonic increasing. Two points labeled by 3 and 4 at the "knees" of the S-shaped curve, are the saddle-node bifurcation points. Therefore, in the range of I_{ext} between the two points, there are three equilibrium points (two stable and one unstable equilibrium points), whereas only one stable equilibrium point exists in the other range. There are also two Hopf bifurcation points labeled by 2 and 5 which locate near the saddle-node bifurcation points.

Similarly to the previous analysis, let us trace periodic solutions which may bifurcate from both Hopf bifurcation points 2 and 5. To do so, clicking on Grab button and using tab key, we grab the Hopf bifurcation (HB) point with label 2. Then, we

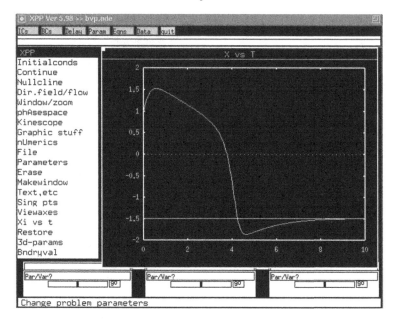

Fig. 3.12 X vs. T plot of the BVP equations (3.1)

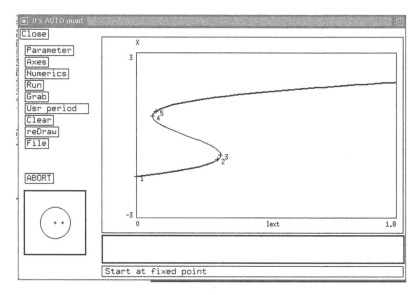

Fig. 3.13 One-parameter bifurcation diagram on steady states of the BVP equations (3.1) with different parameter value

click on Run and Periodic buttons to obtain a branch of periodic orbits bifurcated from the Hopf bifurcation point 2. If we repeat these operations (Grab the point 5, and so on) for another Hopf bifurcation point 5, we can obtain the bifurcation

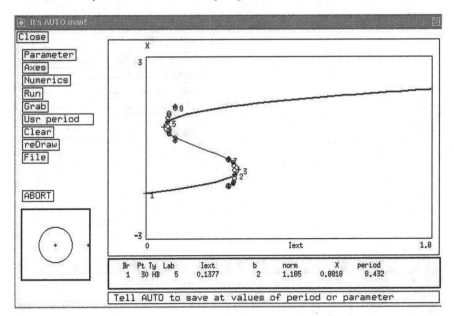

Fig. 3.14 One-parameter bifurcation diagram on *both* steady states and periodic orbits of the BVP equations (3.1)

diagram on both equilibrium points and periodic orbits as shown in Fig. 3.14. Note that the two branches of periodic orbits from two points 2 and 5 are very short. The reason of these short terminations of branches of periodic orbits may be clarified by the following two-parameter bifurcation analysis.

In order to investigate how the locus of the Hopf bifurcation changes when the other parameter value is changed, we make two-parameter bifurcation diagrams. To do so, we Grab the Hopf bifurcation point 2 again. Then we click on Axes and Two par buttons. A new window as shown in Fig. 3.15 will appear, where we change as XMIN=-1, YMIN=-1, XMAX=2, YMAX=4. If we click on Run and Two param buttons, a (part of) two-parameter bifurcation diagram will be obtained as Fig. 3.16, where both axes denote the bifurcation parameters: the abscissa is I_{ext} and the ordinate is another parameter b of the BVP equations (3.1). The short curve is the *Hopf bifurcation curve* which denotes the loci of the Hopf bifurcation. If the two parameters I_{ext} and b are chosen on the Hopf bifurcation curve, then a Hopf bifurcation occurs. Namely, the two-parameter bifurcation diagram illustrates how the locus of the Hopf bifurcation changes with the change of two parameters.

The Hopf bifurcation curve in Fig. 3.16 is not complete. In order to complete the Hopf bifurcation curve, we trace the Hopf bifurcation curve in "another direction." To do so, we click on Numerics button and change the value of Ds to -0.1 as shown in Fig. 3.17. Then, we again Grab the Hopf bifurcation point 2 of Fig. 3.13. Note that, although we cannot see the Hopf bifurcationpoint 2 on the present AUTO

Fig. 3.15 `AutoPlot` window

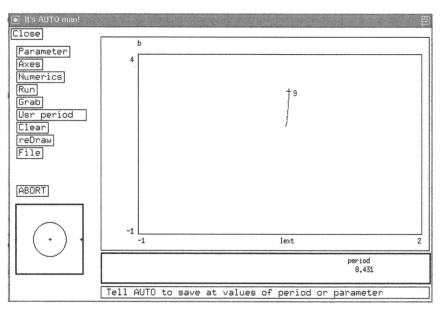

Fig. 3.16 A part of two-parameter bifurcation diagram on Hopf bifurcation

window, we can move to the Hopf bifurcation point (by using `tab` key) *following the information below the bifurcation diagram*, where several information including the Hopf bifurcation label 2, will be shown. Then, if we hit `enter` or `return` key, we can grab the bifurcation point labeled by 2. Finally the clicking on `Run` and `Two param` buttons gives us the whole two-parameter bifurcation curve of Hopf bifurcation as shown in Fig. 3.18.

The bifurcation curve in Fig. 3.18 is the curve traced from the Hopf bifurcation point 2 of Fig. 3.13. Let us draw another bifurcation curve which started from

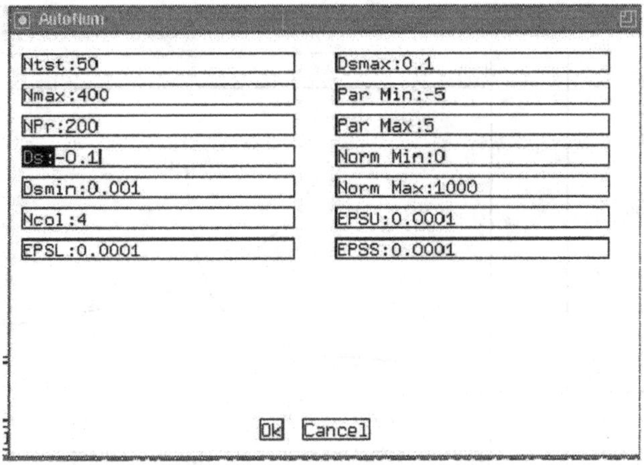

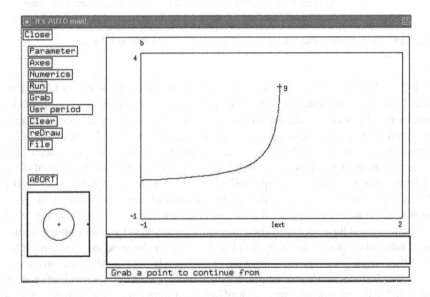

Fig. 3.17 AutoNum window

Fig. 3.18 Whole two-parameter bifurcation diagram on Hopf bifurcation

another Hopf bifurcation point 5 of Fig. 3.13. To do so, we Grab the point 5 following the information below the diagram, similarly to the above case. Then, if we click on Run and Two param buttons, then "a part of another" Hopf bifurcation curve will appear. Completely similarly to the above case, we change the tracing direction by changing the Ds value back to 0.1 (at first, click on Numerics button). Then, if we Grab the Hopf bifurcation point 5 and click on Run and Two param buttons, then whole bifurcation curve of another Hopf bifurcation can be obtained as Fig. 3.19.

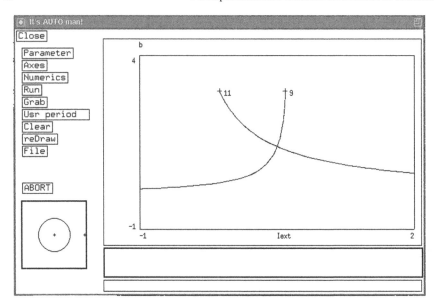

Fig. 3.19 Whole two-parameter bifurcation diagram on "another" Hopf bifurcation

In Fig. 3.19, the two Hopf bifurcation curves terminate at the points labeled by 9 and 11. These terminations are strange, and thus we clarify the reason of the termination by making bifurcation curves of the other bifurcation. Let us make a two-parameter bifurcation diagram on the different bifurcation: the saddle-node bifurcation labeled by 3 in Fig. 3.13. The computational process is completely the same as that of Hopf bifurcation: main ingredients are the twice executions of Grab of point 3 and the "Run + Two param" with changing the tracing directions by the Ds value change. This process gives us the final two parameter bifurcation diagram on both Hopf and saddle-node bifurcations as shown in Fig. 3.20. Note that in order to obtain the saddle-node bifurcation curve, we do not need to Grab another saddle-node bifurcation point labeled by 4 in Fig. 3.13. This is because the saddle-node bifurcation curve is a connected single curve, differently the Hopf bifurcation curves.

In Fig. 3.20, there are special points labeled: the point 14 is a special bifurcation point called the *cusp* point. The two Hopf bifurcation curves terminate on the saddle-node bifurcation curve at two points labeled by 9 and 11 which are also special bifurcation points called the *Bogdanov–Takens* bifurcation point. In both cases of the cusp and Bogdanov–Takens bifurcation points, the equilibrium point of the BVP equations (3.1) possess double zeroes as eigenvalues. However, the behavior of the BVP model may be different between the two special bifurcation points.

Exercise 3.1. Choosing several sets of parameter values of b and I_{ext} on the two-parameter bifurcation diagram of Fig. 3.20, explore the dynamics of the BVP model on x–y phase plane. What is the difference from the BVP model with default parameter values? Consider geometrically what makes these differences if any.   □

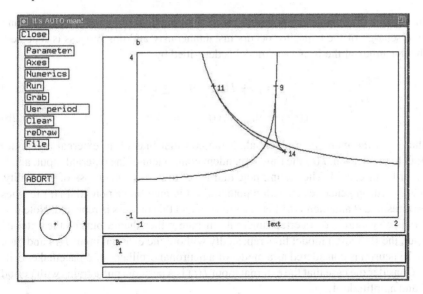

Fig. 3.20 Final two-parameter bifurcation diagram on both Hopf and saddle-node bifurcations

As we have explored so far, one- and two-parameter bifurcation analyses extremely help us to understand the whole behavior of the BVP neuronal model which cannot be obtained by the numerical integrations (simulations) of the BVP equations (3.1) solely. For the more detailed explorations on the dynamics and bifurcations of the BVP or FHN equations, see Braaksma (1993, 1998), Guckenheimer (1986), Koper (1995), Nomura et al. (1994a), and Rocşoreanu et al. (2000).

3.2 Simple Models of Neurons and Neuronal Oscillators

In this section, let us consider very simple or formal models of neurons and neuronal oscillators (pacemakers) and examine the input–output (IO) translation characteristics of these models. Surprisingly, we will show that these simple models arisen in different contexts are all systematically described by a simple one-dimensional discrete-time dynamical system (difference equation) using a concept of the *phase transition curve* (PTC), although the IO characteristics (or response characteristics) of these simple models are very complicated.

3.2.1 Integrate-and-Fire Neuron

Integrate-and-fire (IF) neuronal model is the simplest model in which a neuron is considered to be a single switch with a membrane capacitance and is a special case of the so-called leaky integrator treated in the next subsection (Rescigno et al. 1970; Stein et al. 1972; Glass and Mackey 1979, 1988). The capacitor integrates input

stimuli and if a membrane potential reaches a threshold, the switch is closed and
the discharge of the capacitor occurs instantaneously and this process is repeated.
The dynamics of the IF neuron model is described by

$$\frac{dv}{dt} = \mu + I(t), \quad \text{if } 0 \leq v \leq \theta, \tag{3.2a}$$

$$v(t^+) = 0, \quad \text{if } v(t) = \theta, \tag{3.2b}$$

where v is the membrane potential, μ the constant bias of an external (synaptic)
input to the neuron, $I(t)$ the time-dependent component of the external input, and θ
the firing threshold. The resting potential is set to zero without loss of generality.
The IF neuron generates an action potential if the membrane potential $v(t)$ reaches
the threshold θ and then $v(t^+)$ is reset to zero and this process is repeated. Note that
if μ is considered as a certain internal attribute rather than a bias of the external
input, the IF neuron model fires repeatedly without the external input $I(t)$ and then
the IF neuron is considered as a model of a neuronal oscillator or a pacemaker cell.

Consider the case that the external input $I(t)$ is a periodic pulse trains with period
T and amplitude A:

$$I(t) = \sum_{k=0}^{\infty} A\delta(t - kT), \tag{3.3}$$

where δ is a Dirac's delta function; if a single pulse is injected, the membrane
potential v is increased by A immediately (if v exceeds the threshold θ by the pulse,
the neuron emits an impulse or a spike, and then v is reset to zero):

$$v(t^+) = \begin{cases} v(t) + A & \text{if } 0 \leq v(t) < \theta - A \\ 0 & \text{if } \theta - A \leq v(t) \leq \theta \end{cases} \quad (t = 0, T, 2T, 3T, \ldots).$$

Figure 3.21 shows an example of a phase locking of the IF neuron (3.2) to
the periodic stimulus (3.3). Thick lines in the upper trace shows the value of membrane potential $v(t)$ as a function of t and the lower trace shows the periodic pulsatile

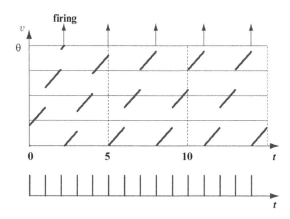

Fig. 3.21 Phase locking
of the integrate-and-fire
neuron model. $\theta = 1$,
$\mu = 0.18$, $A = 0.2$, $T = 1$.
Thick lines of the upper trace
shows the value of membrane
potential $v(t)$ as a function of
t and the *lower trace* shows
the periodic pulsatile stimulus

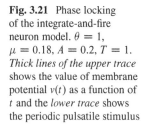

stimulus. The initial value $v(0)$ is zero and the first pulse is injected at $t = 0$; i.e. $v(0^+) = A = 0.2$. The IF neuron generates the first action potential (fires) a short time later the third pulse input ($t \approx 2.2$) and also fires at $t = 5$ by the sixth pulse input. After this firing, the variation of $v(t)$ becomes periodic generating one action potential for every three input pulses: 3:1 phase locking occurs. The mechanism of phase lockings of the IF neuron to pulse stimuli will become clearer if we observe or sample the $v(t)$ value at $t = 0T, 1T, 2T, \dots$, as follows.

Suppose that v_n denotes the value of membrane potential $v(t)$ *immediately after* the n-th pulse injection and v_{n+1} that of $(n + 1)$-th pulse. Then we can obtain the relation between v_{n+1} and v_n as follows:

$$v_{n+1} = p(v_n) \equiv \begin{cases} v_n + A + \mu T \bmod \theta & \text{if } 0 \leq \{(v_n + \mu T) \bmod \theta\} < \theta - A, \\ 0 & \text{if } \theta - A \leq \{(v_n + \mu T) \bmod \theta\} \leq \theta. \end{cases} \tag{3.4}$$

If we otherwise consider v_n as the value of the membrane potential immediately *before* the n-th pulse injection, then the above relation should be replaced by the following:

$$v_{n+1} = p(v_n) \equiv \begin{cases} v_n + A + \mu T \bmod \theta & \text{if } 0 \leq v_n < \theta - A, \\ \mu T \bmod \theta & \text{if } \theta - A \leq v_n \leq \theta. \end{cases} \tag{3.5}$$

If we consider the value of v as the "phase" of the neuron, the maps $p(v)$ of (3.4) and (3.5) describe the transition rule of the phase and thus is called the phase transition curve (PTC) (Kawato and Suzuki 1978; Winfree 1980; Kawato 1981).

In order to study the phase-locking patterns of the IF neuron subjected to periodic pulsatile inputs, it is sufficient to explore the dynamics of the one-dimensional map (circle map) of (3.4) or (3.5). Figure 3.22a,b show the examples of (3.4) and (3.5),

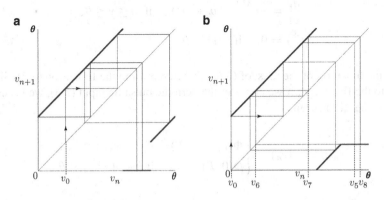

Fig. 3.22 PTC of the IF neuron. $A = 0.2$, $\mu T = 0.18\theta$. Panels (**a**) and (**b**) are different each other only by their coordinates, and correspond to (3.4) and (3.5), respectively

respectively. Thick lines are the graph of $p(v)$ and thin lines show an orbit. Both (a) and (b) have the essentially same dynamics: The difference is just the choice of *coordinate* (the value of $v(t)$ after or before the input). Let us explain the dynamics of the map $p(t)$ using Fig. 3.22b. The orbit $\{v_n\}$ of Fig. 3.22 corresponds to the solution $v(t)$ shown in Fig. 3.21. The initial value v_0 of the orbit in Fig. 3.22b is zero since v_n is the value of $v(t)$ immediately before the pulse injection (see Fig. 3.21). As stated above, the IF neuron fires by the sixth pulse injection at $t = 5T$ which corresponds to v_5 in Fig. 3.22b. After this firing, the orbit becomes periodic: v_6, v_7, v_8, $v_9 = v_6$, $v_{10} = v_7$, ... In Fig. 3.22a, the coordinate or phase v_n is the value of $v(t)$ immediately after a pulse injection and thus the initial value of orbit $\{v_n\}$ of Fig. 3.22a is 0.2 while it is zero in (b). The dynamics of (a) is completely same as (b).

The main part of the maps of Fig. 3.22 is just the shift of v_n by $A + \mu T$ (linear part with a unit slope) and such a linear map hardly produces any phase-locking without $A + \mu T$ being a rational number which is a rare case. The flat (zero-slope) part of the map, however, can produce a phase locking for almost all initial values and parameter values of A and T except for special values in a measure zero set (Doi 1993). Note that a zero slope corresponds to a super stability as stated in Sect. 1.4.1. See the fact that the orbit $\{v_n\}$ becomes periodic after it passes the flat segment in Fig. 3.22. In terms of neurons, the existence of threshold in the IF neuron produces phase lockings.

3.2.2 Leaky Integrate-and-Fire Neuron

Next, we take into account a (linear) resistivity of a membrane in addition to the membrane capacitance of the IF neuron. Then, we obtain a so-called leaky integrate-and-fire (LIF) neuron:

$$\frac{dv}{dt} = -\frac{v(t)}{\tau} + \mu + I(t), \quad \text{if } 0 \le v \le \theta, \tag{3.6a}$$

$$v(t^+) = 0, \quad \text{if } v(t) = \theta. \tag{3.6b}$$

Only the first term of the r.h.s. of (3.6a) is different from the IF neuron (3.2). Similarly to the IF neuron, if we consider the periodic pulsatile input (3.3), we have the PTC of the LIF neuron:

$$v_{n+1} = p(v_n) \equiv \begin{cases} \Phi(v_n + A, T) & \text{if } 0 \le v_n < \theta - A, \\ \Phi(0, T) & \text{if } \theta - A \le v_n \le \theta, \end{cases} \tag{3.7}$$

where $\Phi(x_0, t)$ denotes the solution of the LIF neuron at time t with an initial value x_0 and *without* the periodic stimuli:

$$\Phi(x_0, t) = \begin{cases} (x_0 - \mu\tau)e^{-t/\tau} + \mu\tau & \text{if } 0 \leq t \leq \tau \ln \dfrac{\mu\tau - x_0}{\mu\tau - \theta}, \\ -\mu\tau e^{-t^*/\tau} + \mu\tau & \text{if } t > \tau \ln \dfrac{\mu\tau - x_0}{\mu\tau - \theta}, \end{cases}$$

$$t^* = \left(t - \tau \ln \frac{\mu\tau - x_0}{\mu\tau - \theta}\right) \bmod \left(\tau \ln \frac{\mu\tau}{\mu\tau - \theta}\right).$$

The phase v_n is the value of the membrane potential immediately before the pulse input and we supposed $\mu\tau > \theta$ (the LIF neuron can repetitively fire without the pulse-train stimuli).

Figure 3.23 shows an example of the PTC and its orbit of the LIF neuron. Similarly to the IF neuron, the PTC consists of three linear segments. The different aspect of the LIF neuron from the IF neuron is the slope of the two segments other than the flat (zero-slope) segment. The slope of the left most segment is less than unity which means that the introduction of the membrane resistivity makes the LIF neuron more stable (easier to phase-locked) than the IF neuron. The effect of resistivity, however, is not always the stabilization of the neuron: The slope of the middle segment of the three is *greater* than unity which means that the resistivity sometimes makes a neuronal dynamics more unstable. The LIF neuron, however, is totally very stable and easy to be phase-locked to periodic inputs owing to the flat segment (super stability). See the fact that the orbit in Fig. 3.23 with an initial value $v_0 = 0$ becomes periodic (phase locked) after its passage (v_6) of the flat segment of the map. The flat segment of the PTC is generated by the existence of a firing threshold (and a reset to the resting potential) similarly to the case of IF neuron.

In this section, we have considered the LIF neuron model stimulated by periodic pulse trains. There are, however, vast literatures on the forced LIF neuron: e.g. (to name just a few) the LIF neuron forced by a sinusoidal input (Rescigno et al. 1970; Scharstein 1979; Keener et al. 1981) and the one forced by the more general

Fig. 3.23 PTC of the LIF neuron. $\mu = 0.5, \tau = 4.0$, $A = 0.2, T = 0.5$. Note that, differently from the IF neuron shown in Fig. 3.22, the slopes of the left and middle linear segments are less than and greater than unity, respectively

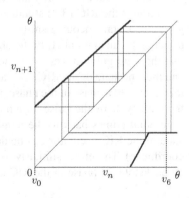

periodic input (Bulsara et al. 1996; Torikai and Saito 1999; Nakano and Saito 2002; Pakdaman 2001). Also see Mirollo and Strogatz (1990) for the synchronization behavior of mutually coupled LIF neuronal oscillators.

3.2.3 Radial Isochron Clock as a Model of Pacemaker Cells

When the parameter μ is considered as an intrinsic aspect of the IF neuron, the IF neuron is considered as a model of pacemaker cells or biological clocks. Such a clock, so to say, corresponds to a sand glass (just an integration or accumulation of a charge). The simplest model which is known as a model of the other type of clocks which correspond to a pendulum, is the so-called radial isochron clock (RIC) or Poincaré oscillator (Guevara and Glass 1982; Hoppensteadt and Keener 1982; Keener and Glass 1984; Glass and Mackey 1988; Glass and Sun 1994; Nomura et al. 1994b):

$$\frac{dr}{dt} = ar(1 - r), \quad a > 0 \tag{3.8a}$$

$$\frac{d\theta}{dt} = 2\pi \tag{3.8b}$$

which are expressed in the polar coordinate. These equations possess a unique limit cycle with a radius $r = 1$ and with a unit period. Figure 3.24a shows the phase plane of (3.8) in *Cartesian coordinate*. (Do not confuse the "phase" of the phase plane with the phase of the above subsections.) The unit circle is the stable limit cycle and solution curves started from any points of the phase plane (except for the origin) asymptotically approach the limit cycle. The parameter a controls the stability of the limit cycle (the strength of the attraction of the solution curves by the limit cycle). The reason that this oscillator is called the RIC is that its isochrons (a set of state points in the phase plane which asymptotically converge to the same point on the limit cycle) are radial (Guckenheimer 1975).

Consider the RIC (3.8) stimulated by a periodic pulse trains (3.3) where a single pulse instantaneously shifts the state point in Fig. 3.24a in a right horizontal direction by an amount A. In the absence of such external stimuli, we can consider that the state point moves in the neighborhood of the stable limit cycle (attractor) and thus the state of the RIC oscillator can be denoted by a phase θ (do not confuse this term with a phase of the phase plane). If a pulse is injected, the state point leaves the limit cycle by the pulse for a moment. After an enough long time, however, the state point comes back to the neighborhood of the limit cycle again and thus the state of the RIC oscillator can be denoted by the phase θ (the angle θ in the polar coordinate). Therefore, similarly to the case of the IF neuron with periodic stimuli, let us denote the phase of the RIC immediately before the n-th pulse by θ_n and that

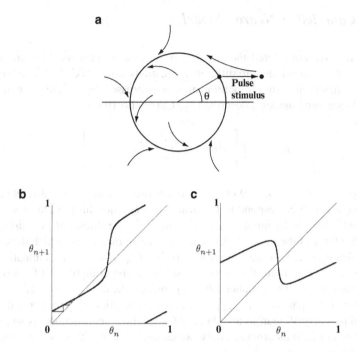

Fig. 3.24 (**a**) Phase plane of the radial isochron clock (RIC). (**b, c**) PTC of the RIC. Parameter values are: (**b**) $A = 0.9$, $T = 0.1$; (**c**) $A = 1.1$, $T = 0.5$

of $(n + 1)$-th pulse by θ_{n+1}. From a simple calculation, we can obtain the relation (PTC of the RIC) between θ_{n+1} and θ_n as follows:

$$\theta_{n+1} = p(\theta_n) \equiv \arctan \frac{\sin \theta_n}{A + \cos \theta_n} + 2\pi T \mod 2\pi. \tag{3.9}$$

Figure 3.24b,c show the examples of the PTC $p(\theta)$ of the RIC. Panel (b) corresponds the case that the amplitude of the external pulse is less than the radius of the limit cycle ($A < 1$) and (c) the case of $A > 1$. The map of (b) and (c) are continuous maps on S with degree unity and zero, respectively (see Sect. 1.4.1). The phase locking of the RIC can be studied by examining the one-dimensional map (3.9) similarly to the case of the IF neuron. The map of (b) has the similar dynamics to that of the sine-circle map (1.46) with small b values. The map of (c) has a typical feature as a bimodal map with two increasing segments and one decreasing segment, and the characteristics of this bimodal map or the RIC will be clarified later. The essential difference between the phase-locking behavior of the RIC and that of the IF neuron is that RIC can produce chaotic behavior which is not appeared in the IF neuron (we can easily expect chaotic behavior in the RIC from the bimodal feature of PTC $p(\theta)$ of the RIC).

3.2.4 Caianiello's Neuron Model

So far, we have considered the simple neuron models expressed by differential equations and reduced them to difference equations using PTC. In this subsection, we will consider the simple neuron models expressed by difference equations. Caianiello's neuron model is described by (Caianiello 1961):

$$x_{n+1} = \mathbf{1}\left[S(n) - \alpha \sum_{k=0}^{n} b^{-k} x_{n-k} - \theta \right], \qquad (3.10)$$

where $\mathbf{1}[x] = 1 \ (x \geq 0), = 0 \ (x < 0)$. The neuron's state x_n at a discrete time n takes 1 or 0 which correspond to the firing state and non-firing state of a neuron, respectively. $S(t)$ is an input to a neuron and θ is the threshold of a firing. The summation term of the r.h.s. of (3.10) denotes the refractoriness which depends on the past firing $\{x_{n-k}\}$. If the neuron fired before $(x_{n-k} = 1)$, the summation term becomes negative and thus the neuron must have a bigger input $S(n)$ for a firing at the next time $n + 1$. The parameter $b(> 1)$ controls the decay rate of refractoriness depending on the past state $\{x_{n-k}\}$. The unit step function $\mathbf{1}[x]$ corresponds to the threshold property of neurons and thus the Caianiello's neuron model is an abstract neuron model which incorporates only two characteristics of threshold property and refractoriness of a neuron.

Nagumo and Sato (1972) investigated the Caianiello's neuron model in detail. When a constant input $(S(n) \equiv A)$ is applied to the neuron, changing the variable x_n to y_n as follows:

$$y_n = (A - \theta)/\alpha - \sum_{k=0}^{n} b^{-k} x_{n-k} \qquad (3.11)$$

(3.10) is reduced to

$$y_{n+1} = p(y_n) \equiv y_n/b - \mathbf{1}[y_n] + a, \qquad (3.12)$$

where we have set as $a \equiv (A - \theta)(1 - 1/b)/\alpha$. Again, we have obtained a one-dimensional map $p(y)$ of (3.12). Figure 3.25a is an example of the map $p(y)$ which is a piecewise linear map with two increasing segments with a slope $1/b$. Whole dynamics of such a piecewise linear map is well known (Leonov 1959; Nagumo and Sato 1972; Hata 1982): the map has a stable periodic orbit for almost all parameter values (except for a measure zero set). As is seen from the definition of the variable y_n, the right branch $(y_n > 0)$ and the left branch $(y_n < 0)$ correspond to a firing and a non-firing, respectively. In Fig. 3.25a, an orbit $\{y_n\}$ with $y_0 = -1$ is also drawn. This orbit asymptotically converges to a periodic orbit $\{y_1^*, y_2^*, y_3^*, y_4^*, y_5^*, y_1^*, y_2^*, \ldots\}$ which corresponds to a firing pattern 10101 since y_1^*, y_3^* and y_5^* pass the right branch and y_2^* and y_4^* pass the left branch (where a firing and non-firing are denoted by the symbols "1" and "0", respectively).

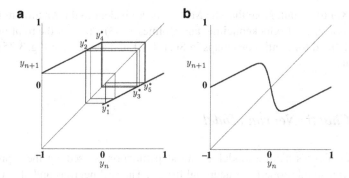

Fig. 3.25 (a) Caianiello's neuron model ($a = 0.7$, $b = 2$). (b) Chaos neuron model ($\epsilon = 0.05$, $a = 0.5$, $b = 2$)

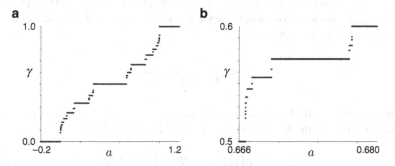

Fig. 3.26 Mean firing rate (MFR) of the Caianiello's neuron model (3.12) as a function of the parameter a. The other parameter is set as $b = 2$. (b) is the magnification of a portion of (a): a similar structure to (a) is nested between the steps of (a). This nesting structure (fractals) continues infinitely and constitutes a Cantor function or devil's staircase (zero differential coefficient almost everywhere, non-decreasing and continuous function)

Let us define a mean firing rate (MFR) for the solution orbit $\{y_n\}$ of the one-dimensional map (3.12) as follows:

$$\gamma = \lim_{n \to +\infty} \frac{1}{n} \sum_{k=0}^{n-1} \mathbf{1}[y_k]. \tag{3.13}$$

The MFR of Fig. 3.25a is 3/5 since its firing pattern is 10101. Figure 3.26 shows the MFR γ as a function of the parameter a which corresponds to the amplitude A of the input to the Caianiello's neuron. Panel (b) is the magnification of a portion of (a); between the steps or stairs of (a), many small stairs are nested and this nesting structure continues infinitely. The complicated structure (fractals) of the MFR is mathematically a Cantor function (zero slope almost everywhere, non-decreasing and continuous function) and also is called a *devil's staircase*. Mathematical verification of this complicated structure is not so difficult, see Leonov (1959) or

Mira (1987) for detail. Note that the MFR γ is equivalent to the rotation number ρ defined in Sect. 1.4.1 (Guckenheimer and Holmes 1983) although the rotation number was defined for continuous maps in Sect. 1.4.1 and the map of Fig. 3.25a is not continuous.

3.2.5 Chaotic Neuron Model

In the Caianiello's neuron model, a unit step function is used for the representation of a threshold property of neuronal firings. The real neurons and also realistic neuronal models such as the Hodgkin–Huxley equations, however, do not possess such a rigorous threshold (i.e. not all-or-none) and do possess intermediate firing states (neither firing nor non-firing). Aihara et al. (1990) incorporate this property into the Caianiello's neuron model. They replaced the step function by the following sigmoidal (or logistic) function:

$$f(x) = \frac{1}{1 + \exp(-x/\epsilon)}. \tag{3.14}$$

In the limit of $\epsilon \to 0$, this function converges to the step function. Similarly to the Caianiello's neuron model, we can obtain a one-dimensional map:

$$y_{n+1} = p(y_n) \equiv y_n/b - f(y_n) + a. \tag{3.15}$$

Figure 3.25b shows an example of the graph of $p(y)$. The Caianiello's neuron model of Fig. 3.25a has two discontinuous segments. In the map of Fig. 3.25b, the two segments are connected continuously and the map has a typical bi-modal structure and thus can produce chaotic behavior differently from the Caianiello's neuron model. Thus, this neuron model is called the chaotic neuron model. (Real neurons and the HH equations also present a chaotic behavior, Aihara et al. 1984; Matsumoto et al. 1984; Takahashi et al. 1990.) For the chaotic neuron model, we can (numerically) calculate the MFR similarly to the Caianiello's neuron model (3.10). The graph of MFRs of the chaotic neuron model, however, does not constitute the complete devil's staircase but does constitute the partial or incomplete staircase owing to the difference of the topological feature of the two maps shown in Fig. 3.25a,b (Yellin and Rabinovitch 2003).

Comparing Figs. 3.25b and 3.24c, we can see the similarity of topological features of the two graphs although their functional forms are different. Thus, it is apparent that two neuronal models (RIC and chaos neuron model) arisen in different contexts and with different intrinsic dynamics have the essentially same phase-locking or input–output characteristics. This is a very advantage of the usage of such simplified or abstract models.

3.3 A Variant of the BVP Neuron Model and Its Response to Periodic Input

3.3.1 Piecewise-Linearized BVP Model

In the following, we consider a variant of the BVP model (3.1):

$$\epsilon \dot{x} = f(x, y) \equiv y - x + 5/6\{|x + 1| - |x - 1|\}, \tag{3.16a}$$
$$\dot{y} = g(x) \equiv -x + a, \tag{3.16b}$$

where the cubic polynomial in the original BVP model (3.1a) is approximated by a piecewise linear function and the parameters are also simplified. Note that the externally applied constant current I_{ext} of (3.1) is now considered to be included in the new parameter a of (3.16) (by a change of variables). Also, note that the sign of the variable x in (3.16) is reversed from the previous BVP model (3.1); the negative direction of x or the left direction on the x–y phase plane corresponds to the depolarization or the excitation of a neuron. We consider that this simplified BVP model still preserves the neuronal features of the original BVP model and of the HH model although this equation is much simpler than (3.1). Throughout the following sections of this chapter, (3.16) will be referred to as the BVP model unless otherwise specified in the text.

Let us explain the "neuronal" features of the BVP model (3.16). Figure 3.27 shows the x–y phase plane of (3.16). The N-shaped line is the x-nullcline on which $\dot{x} = 0$ and orbits (solutions) of (3.16) move vertically, and the vertical line is the y-nullcline ($\dot{y} = 0$). So the intersection point P of the two nullclines is an equilibrium or a fixed point of (3.16) which corresponds to the resting state or the quiescent state of a neuron. Typical orbits $(x(t), y(t))$ with different initial values of (3.16)

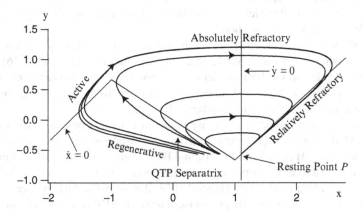

Fig. 3.27 The x–y phase plane of (3.16) with the N-shaped x-nullcline ($\dot{x} = 0$) and the linear y-nullcline ($\dot{y} = 0$). The parameters are $\epsilon = 0.1$, $a = 1.1$

are also displayed. If the neuron is in its resting state and a pulse-like current is injected, the neuron responds and settles down back to the resting state. For example, an orbit starts at the equilibrium or the resting point P, and a large single-pulse perturbation in the negative (depolarizing) direction of the x axis displaces it leftwards, it goes through both the regenerative and active regions, and finally comes back to P. (Note that, in the modified BVP equations (3.16), the negative direction of x corresponds to the depolarization or the action potential differently from the original BVP equations (3.1).) If the perturbation is small, the orbit does not enter the regenerative region, i.e. the BVP neuron model is not excited. A phase-plane analysis is very useful so that we can get the intuitive understanding of the total behavior of the model without need to make brute-force computer simulations of the model with many different initial values.

The response of the BVP model to a single pulse perturbation is not all-or-none (living neurons are not all-or-none also). There are neutral orbits which we cannot decide whether they excite or not. The orbits around QTP (quasi-type) separatrix are such type's orbits. In this region, the orbit changes its amplitude drastically (but continuously) depending on its initial value. Thus, the QTP separatrix is considered as a (quasi-) threshold of the neuron model. Discussions on the geometric property of this threshold curve and on the excitability of the BVP neuron model (3.1) are made by Okuda (1981). The time waveforms "$-x(t)$" of the orbits shown in Fig. 3.27 are plotted in Fig. 3.28a. Depending on the strength of the input pulse current, various amplitude of responses are generated.

Figure 3.28b shows a response characteristics of the BVP neuron model to a single pulse stimulus. The abscissa shows the amplitude of such a depolarizing stimulus, and the ordinate shows the response amplitude (the maximum values of "$-x(t)$") of the neuron model. In the case of $\epsilon = 0.1$ (solid curve), the response amplitude grows rapidly in the range [0.3,0.4] of the abscissa. So the BVP model is considered to have a certain (quasi-) threshold property. If the parameter ϵ decreases to 0.05 (dashed curve), this threshold property becomes more strict; the response amplitude changes abruptly (but still continuously) near a value 0.28. In the limit case of $\epsilon = 0$, the BVP model responds ideally (all-or-none) and the response characteristics becomes a step function.

Let $(x^*, y^*) = (a, a - 5/3)$ denote the resting point P. This equilibrium point is stable when $|a| > 1$ and unstable when $|a| < 1$. In the latter case, (3.16) has a stable limit cycle and the BVP model excites repetitively. So, we can consider the BVP model (3.16) with $|a| > 1$ a neuron (excitable cell) model and the one with $|a| < 1$ a neuronal oscillator (pacemaker cell) model.

A neuron shows a repetitive activity when an external constant current is applied to. Mathematically this repetitive activity corresponds to a limit cycle solution of a model. The HH model and the original BVP model (3.1) have such a limit cycle which is "bifurcated" from an equilibrium point (a resting state) by the Hopf bifurcation (Hadeler et al. 1976; Hassard 1978; Rinzel 1978). Strictly speaking, no such a bifurcation occurs in the piecewise-linearized BVP model (3.16) since the right hand side (r.h.s) of (3.16) at the equilibrium point P is not differentiable at the critical parameter value $a = 1$. This difference between (3.1) and (3.16) does not

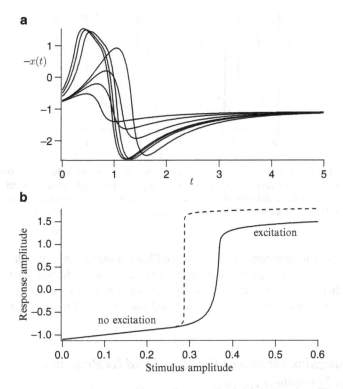

Fig. 3.28 (a) Response waveforms $-x(t)$ to various amplitude of single pulses. The time wave-forms of various orbits which are shown in Fig. 3.27 are plotted. (b) Response characteristics of the BVP model (3.16) to a single pulse input, with parameter values $\epsilon = 0.1$ (*solid curve*) and $\epsilon = 0.05$ (*dashed curve*). The amplitudes (the maximum values) of $-x(t)$ are plotted against the amplitudes of single pulse inputs

affect the total behavior of the model: The model (3.16) has a stable equilibrium point when $|a| > 1$ and has both an unstable equilibrium point and a stable limit cycle when $|a| < 1$, although a slight delicate phenomenon occurs at the critical value $|a| = 1$. The amplitude of the limit cycle grows as the value of $|a|$ decreases away from unity. This qualitative behavior of (3.16) is the same as that of (3.1).

Figure 3.29 shows a repetitive activity of the BVP model with different parameter values: $a = 0.0$ (dashed curve) and $a = 0.9$ (solid curve). As the value of a increases, the asymmetry of the waveform increases and the depolarized $(-x(t) > 0)$ fraction of time in one cycle (period) decreases. This fraction of time is called an *activity* of a neuron model and the importance of asymmetry is already noted in the context of coupled neuronal oscillators (Kepler et al. 1990; Meunier 1992). In the figure, the parameter ϵ is set as much smaller than one used in Figs. 3.27 and 3.28. As the value of ϵ approaches zero, the waveform of $-x(t)$ loses its smoothness.

As stated above, the parameter a may be considered as an "external" current applied to a neuron. If we regard a as an "intrinsic" property of a neuron, then the BVP

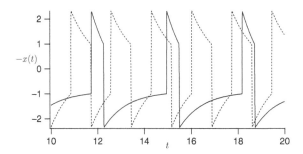

Fig. 3.29 Repetitive activities of the BVP model (3.16) with parameter values: $a = 0.0$ (*dashed curve*) and $a = 0.9$ (*solid curve*). The parameter ϵ is set to 0.001 which is much smaller than one used in Fig. 3.27 and the waveforms have a sharp shape. As the a value increases, the fraction time such that $-x(t) > 0$ decreases and the waveform becomes asymmetric (neuron-like)

model (3.16) with a parameter $|a| < 1$ oscillates spontaneously and thus we can consider the neuron as a neuronal oscillator or a pacemaker cell. Note that the BVP model (3.16) with $a = 0$ is equal to the piecewise-linearized version of the van-der Pol oscillator (van der Pol 1926; Grasman and Jansen 1979; Xu and Jiang 1996).

3.3.2 Singular Version of the Model and Its Response to Sinusoidal Input

In the following, we consider the BVP model (3.16) in the limit of $\epsilon = 0$. Figure 3.30 shows the x–y phase plane of (3.16) with $\epsilon = 0$. The parameter is set as $|a| < 1$, so the equilibrium point is unstable and the model shows a repetitive activity. The rectangle ABCD is a stable limit cycle.

In the limit of $\epsilon = 0$, the orbit with an initial value which is not on the x-nullcline ($\dot{x} = 0$) *instantaneously* jumps to the x-nullcline in the horizontal direction since the velocity $\dot{x} = f(x, y)/\epsilon$ becomes infinity unless $f(x, y) = 0$. Thus, all orbits are considered to move on the x-nullcline. On the right (left) branch of the N-shaped x-nullcline, an orbit moves toward the point B (D, resp.). At the point B (D), the orbit instantaneously jumps to the point C (A, resp.), since we consider the limit of $\epsilon = 0$. This limit case is called a *singular limit* since orbits are not differentiable at such a jump point. Note that all orbits move on the x-nullcline even in the case of a non-oscillatory parameter value $|a| > 1$ although all orbits tend to the equilibrium (resting) point after a suitable time in this case.

Let us introduce a sinusoidal external input to the BVP model (3.16):

$$\epsilon \dot{x} = f(x, y) = y - x + 5/6\{|x + 1| - |x - 1|\}, \tag{3.17a}$$

$$\dot{y} = g(x) + v(t) \equiv -x + a + A\sin\{2\pi(t/T + \theta_0)\}, \tag{3.17b}$$

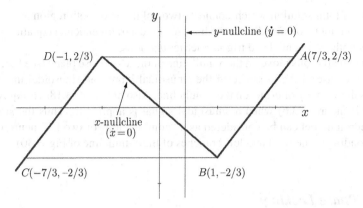

Fig. 3.30 The x–y phase plane of (3.16) in the limit of $\epsilon = 0$ and the N-shaped x-nullcline ($\dot{x} = 0$). The closed orbit ABCD is a limit cycle. The parameter a shifts the y-nullcline in the horizontal direction ($a = 0.4$ in this figure)

where A, T, θ_0 are the amplitude, the period, the initial phase of the sinusoidal input, respectively.

Note that the external stimulus $v(t)$ is applied to the variable y rather than x. It is usual to apply the stimulus to the variable x since x is considered as a voltage and the dimension of the r.h.s. of (3.17a) corresponds to a current. Mathematically, however, the application of a current stimulus to y is equivalent to the one to x as follows (Alexander et al. 1990): If $(x(t), y(t))$ is a solution of (3.17) and define $\xi(t) \equiv x(t)$ and $\eta(t) \equiv y(t) - \int v(t)dt$, then $(\xi(t), \eta(t))$ is a solution of

$$\epsilon\dot{\xi} = f(\xi, \eta) + \int v(t)dt,$$

$$\dot{\eta} = g(\xi).$$

Thus the addition of $v(t)$ to the r.h.s. of (3.17b) is equal to that of $\int v(t)dt$ to the r.h.s. of (3.17a).

In the singular limit, (3.17) is reduced to

$$\dot{y} = g(x) + v(t), \quad f(x, y) = 0$$

or more explicitly, to

$$\dot{x} = \dot{y} = -x + a + A\sin\{2\pi(t/T + \theta_0)\}, \tag{3.18a}$$

$$y = \begin{cases} x - 5/3 & (x \geq 1), \\ x + 5/3 & (x \leq -1). \end{cases} \tag{3.18b}$$

This differential equation is a one-dimensional (piecewise) linear equation. Thus we can obtain its analytical solution piecewise (in each region $x \geq 1$ or $x \leq -1$).

However, a total solution which connects two solutions of both region $x \geq 1$ and $x \leq -1$ can not be obtained since we have to solve a transcendental equation which includes both exponential and trigonometric functions.

Note that all orbits move on the x-nullcline of the $x-y$ phase plane (see Fig. 3.30) even in the case of the presence of the sinusoidal input; the sinusoidal input $v(t)$ changes the velocity of orbits on the x-nullcline. Also note that (3.18a) is equivalent to the LIF neuron (3.6) when a sinusoidal signal is applied to. Thus the singular BVP neuron model can be considered as a "combination" of two LIF neurons (i.e. corresponding to the right and left branches of the x-nullcline of Fig. 3.30).

3.3.3 Phase Lockings

Figure 3.31 shows examples of responses of the BVP model (3.17) with $\epsilon = 0.001$ to a sinusoid. Panel (a) corresponds to a neuronal oscillator ($a = 0.9$) and (b)–(d) a (non-oscillatory) neuron ($a = 1.1$). In (a) and (b), the neuron model excites once during one period of a sinusoid. This response pattern is called a 1:1 phase locking. Generally, a response in which m cycles of an input correspond to n spikes of a neuron is called an *m:n phase locking*. Panels (c) and (d) show a 2:1 phase locking and a 1:2 phase locking, respectively. Note that we do not see any essential difference between an oscillatory neuron in (a) and a non-oscillatory neuron in (b).

In order to study the relation between the sinusoidal forcing and the forced BVP model, let us observe the sinusoidal input whenever the orbit (x, y) visits the point A of the $x-y$ phase plane (see Fig. 3.30). If the parameter a is set in the range of $|a| > 1$ (non-oscillatory neuron), orbits will stay near the resting point P and may not visit the point A without any input $v(t)$. If the amplitude of $v(t)$ is sufficiently large, an orbit can excite (move to point C) and then visits the point A even in the case of non-oscillatory neuron.

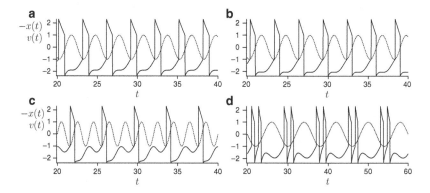

Fig. 3.31 Examples of phase lockings of the forced BVP model (3.17). The amplitude A of the sinusoidal input $v(t)$ is set to 1.0 and other parameters are: (**a**) $a = 0.9$, $T = 3.0$ (**b**) $a = 1.1$, $T = 3.0$ (**c**) $a = 1.1$, $T = 2.0$ (**d**) $a = 1.1$, $T = 8.0$. The *solid curves* are waveforms of $-x(t)$ and *dashed curves* are that of $v(t)$

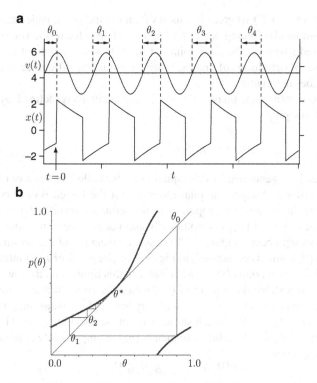

Fig. 3.32 (**a**) Phase of the sinusoidal input (a 1:1 phase locking). Waveforms of both $x(t)$ and the input $v(t)$ are plotted ($v(t)$ is shifted upward and its scale is arbitrary). $A = 0.3$, $T = 1.5$. The peaks (maximum values) of $x(t)$ corresponds to the point A of the $x-y$ phase plane (Fig. 3.30). At the time $t = 0$, the initial phase of the sinusoidal input is θ_0 and the phase is changed to θ_1, $\theta_2, \ldots$ as the time elapses. In this case, the sequence of the phases converges to a certain fixed value θ^*; a 1:1 phase locking occurs. (**b**) Phase transition curve (PTC) $p(\theta)$ (*thick curve*) and its orbit $\{\theta_n\}$ (*thin lines*). $A = 0.3$, $T = 1.5$. The orbit $\theta_0, \theta_1, \ldots$ asymptotically converges to a fixed point θ^* which corresponds to a 1:1 phase locking of (a)

Figure 3.32a is an example of waveforms of both $v(t)$ and $x(t)$. The point A of the $x-y$ phase plane corresponds to the peaks of the waveform $x(t)$. The phase of the sinusoidal input (at the point A) is a number between 0 and 1 normalized by the period T. Note that this terminology "phase" is irrelevant to the "phase" of the $x-y$ phase plane. Suppose that the orbit starts at the point A of the $x-y$ phase plane with an initial phase θ_0 of the sinusoidal input and that the orbit returns to the point A again after a time t_1. Then the second phase θ_1 at the second visit to A is obtained by

$$\theta_1 = p(\theta_0) \stackrel{\text{def}}{=} \theta_0 + t_1/T \quad (\text{mod } 1). \qquad (3.19)$$

The relation $p(\theta_0)$ between θ_0 and θ_1 is called a return map or a phase transition curve (PTC). In Sect. 3.2, the phase of a forced neuron was defined using *the state of the neuron* when the state of the forcing was fixed (i.e. the membrane potential was sampled when $t = kT$) whereas in this BVP model the phase is defined using the state of the sinusoidal forcing when the neuron's state is fixed (the sinusoidal forcing is observed whenever the state point $(x(t), y(t))$ passes the point A). Note that this difference of definitions of phases, however, is just the difference of coordinate choice and does not matter at all.

Using the return map, phases θ_2, θ_3, ... are recursively defined by the one-dimensional mapping:

$$\theta_{n+1} = p(\theta_n), \quad n = 0, 1, 2, \ldots. \tag{3.20}$$

The sequence $\{\theta_n\}$ generated by this equation is also called the orbit of (3.20).

Thus, in order to analyze the phase lockings of the forced BVP model (3.17), we investigate the asymptotic properties of the sequence (orbit) $\{\theta_n\}$ of phases. In Fig. 3.32a, the sequence $\{\theta_n\}$ asymptotically approaches a certain values θ^*; in this case a 1:1 locking occurs. Figure 3.32b shows an example of the return map $p(\theta)$ and its orbit $\{\theta_n\}$ which corresponds to Fig. 3.32a. The graph of $p(\theta)$ intersects with a diagonal line in two points (indistinguishable in this figure) and the lower intersection point θ^* is a stable fixed point of (3.20) (see the fold bifurcation described in Sect. 1.4.1). Thus an orbit θ_0, θ_1, ..., with any initial phase θ_0 asymptotically converges to this fixed point θ^*, which shows a 1:1 phase locking occurs. Generally, in the case of m:n locking, an orbit or sequence $\{\theta_n\}$ asymptotically approaches to an n-periodic sequence:

$$\theta^{(1)}, \theta^{(2)}, \ldots, \theta^{(n)}, \theta^{(1)}, \ldots.$$

Note that we can not obtain the explicit form of $p(\theta)$ analytically by the same reason as stated in Sect. 3.3.2. So, we numerically calculated $p(\theta)$ solving some transcendental equations by the *Newton method*. The computation time of $p(\theta)$ is very short since the convergence of the Newton method is much faster than the numerical simulation of differential equations.

3.3.4 Bifurcation Diagrams

Patterns of phase lockings depend on both the amplitude and the period of the sinusoidal input. One phase-locking pattern can be considered to be bifurcated from the other phase-locking pattern as a (bifurcation) parameter is changed. A bifurcation diagram is a very useful tool to systematically investigate the effect of a parameter change on a system's behavior.

Figure 3.33 shows how the asymptotic value(s) of the sequence $\{\theta_n\}$ is changed depending on the input period T. Panel (a) is a bifurcation diagram of $a = 0.0$ (original van der Pol oscillator). For example, in the $T = 2$ case, only one dot is plotted in the vertical direction ($\theta^* \approx 0.14$); a 1:1 locking occurs. In the $T = 7$ case, three dots are plotted and a 1:3 locking occurs; one cycle of the sinusoidal input synchronizes with three cycles of the oscillator. In the $T = 4$ case, many points

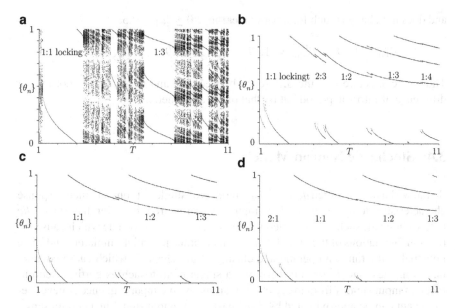

Fig. 3.33 Bifurcation diagrams of the return map or the PTC $p(\theta)$. $A = 1.0$. A stationary (or asymptotic) sequence after some transient initial sequence $\{\theta_n\}$, $n = 101, \dots, 600$ produced by (3.20) were plotted for each of 600 equally spaced T values on the interval $[1, 11]$. (**a**) $a = 0$ (the van der Pol oscillator) (**b**) $a = 0.5$ (the BVP oscillator) (**c**) $a = 0.9$ (the BVP oscillator) (**d**) $a = 1.1$ (non-oscillatory neuron)

are plotted; no locking occurs (in this case, the sequence $\{\theta_n\}$ approaches a so-called quasi-periodic sequence rather than a periodic sequence). In a quasi-periodic case, the oscillator is not phase-locked to the input and the phase of the input takes infinitely many values between 0 and 1 although the oscillator seems to be locked for a short time range (quasi-periodic orbits are different from chaotic ones).

In panel (b), the value of a is set to 0.5. As stated in Sect. 3.3.1, the asymmetry of the waveform increases if the absolute value of a increases. The larger $|a|$ becomes, the narrower the T range of quasi-periodic orbits becomes, although the total bifurcation structure (1:1, 1:3 lockings of (a)) does not change much. From this observation, we can see that the asymmetry of an oscillator increases the tendency to phase lockings. In panels (c) and (d), the value $|a|$ is increased more. We can still see the similarity in all the figures (a)–(d) although the bifurcation diagrams seem to be prolonged to the right direction as $|a|$ increases. A notable point is that we can see no essential difference between (c) and (d) although the panel (c) corresponds to a neuronal oscillator while the panel (d) a non-oscillatory neuron.

We note that the intrinsic period T_a^{Int} of the singular BVP model which has a repetitive activity ($|a| < 1$) without the sinusoidal input ($v(t) \equiv 0$) is given by

$$T_a^{\mathrm{Int}} = \ln \frac{(7/3)^2 - a^2}{1 - a^2}$$

and does not change much for a parameter range $0 \leq |a| \leq 0.5$:

$$T_{0.0}^{\text{Int}} = 1.69, \quad T_{\pm 0.3}^{\text{Int}} = 1.77, \quad T_{\pm 0.5}^{\text{Int}} = 1.94, \quad T_{\pm 0.9}^{\text{Int}} = 3.19.$$

Thus we conjecture that the difference between (a) and (b) is caused not by the difference of intrinsic period but by that of the asymmetry.

3.4 Stochastic Neuron Models

So far, we have considered *deterministic* models of single neurons and their response characteristics to also deterministic inputs. A real neuron, however, has *stochastic* natures in itself, such as spontaneous releases of a synaptic vesicle which cause the random fluctuations of the order 0.5 mV in membrane potential (miniature end-plate potentials), the random opening and closing of ion channels which cause conductance changes, and so on. Furthermore, a single neuron such as cortical neurons receives input signals from many other neurons and the input sequence is never periodic but can be approximated by a stochastic or random signal in a certain sense. (For the stochastic nature of neurons and the stochastic neuron models, see Ricciardi 1977 and Tuckwell 1988, 1989.)

In this section we briefly summarize the input–output characteristics of the IF neuron and the LIF neuron to a Gaussian white noise input with noise intensity σ. The noise is interpreted as a random component of the total inputs to a single neuron or as an internal fluctuations as stated above.

In the following, we focus on the mean *interspike interval* (ISI) and the *coefficient of variation* (CV) of the ISIs as functions of the input parameters, the constant bias μ and the noise intensity σ. Let us denote the ISI as a random variable T in this section. Then, the mean $E[T]$ of the ISI is the inverse of the mean firing rate, and the CV, which describes the relative width of the ISI histogram, is defined as follows:

$$\text{CV} \equiv \frac{\sqrt{\text{Var}[T]}}{E[T]}.$$

The dimensionless CV is often used as a measure of spike train irregularity. For a very regular spike train (pacemaker), the ISI histogram will have a very narrow peak and $\text{CV} \approx 0$. When a neuron fires periodically with period 2 and the observed series of ISIs are $\{5, 10, 5, 10, 5, 10, \ldots\}$, the resultant CV is $1/3$. If a spike train is a completely random process (i.e. a *Poisson process*), the corresponding ISI histogram is an *exponential distribution* and $\text{CV} = 1$.

After the report of Softky and Koch (1993), a high CV at high firing rates observed in the cortex has led to numerous speculations on the nature of the neural code (Shadlen and Newsome 1994, 1998; Konig et al. 1996).

3.4.1 Stochastic Integrate-and-Fire Neuron

Let the time-variable external input $I(t)$ in the IF neuron (3.2) be the Gaussian white noise with noise intensity σ. Then, the time variation of the membrane potential $V(t)$ of a neuron becomes the well-known Wiener process and is described by a stochastic differential equation as follows:

$$dV(t) = \mu dt + \sigma dW(t), \tag{3.21}$$

where $W(t)$ is the standard *Wiener process* or the *Brownian motion* (Gardiner 1983). The value of μ is the bias of the total sum of many excitatory and inhibitory inputs and the term $\sigma dW(t)$ is the (random) deviation from the bias. This stochastic IF neuron has been originally proposed by Gerstein and Mandelbrot (1964) as an approximation of the random walk model.

The ISI is represented by the random variable

$$T_\theta = \inf\{t \geq 0 : V(t) \geq \theta\}, \quad V(0) = 0 < \theta \tag{3.22}$$

which is the *first-passage time* (FPT) through θ of the stochastic process $V(t)$. In this framework, the ISIs are generated in accordance with a renewal process and therefore the nature of the ISI sequence is completely described by the probability density function (pdf) of the T_θ which is the theoretical counterpart of the ISI histogram. For the Wiener process with a constant threshold, the distribution of T_θ is the well-known *Inverse Gaussian Distribution* (IGD) (see Chhikara and Folks 1988 for details of the distribution) and the pdf is as follows:

$$g(t) = \frac{\theta}{\sqrt{2\pi\sigma^2 t^3}} \exp\left[-\frac{(\theta - \mu t)^2}{2\sigma^2 t}\right], \quad t > 0. \tag{3.23}$$

When the constant bias μ is zero or toward the threshold θ ($\mu > 0$), an action potential is generated in a finite time with probability one. On the other hand, when it is away from the threshold ($\mu < 0$), there is a probability that no action potential is ever generated. Also, for $\mu = 0$ which is the case that a neuron gets completely same amount of excitatory and inhibitory inputs, the mean ISI becomes infinite. Therefore only the IF neuron with the positive constant bias μ is adequate for a model of active neurons. Figure 3.34 shows the examples of ISI densities for the IF neuron with various values of the parameters. From this figure, it is evident that the ISI densities of the stochastic IF neuron can have a wide variety of shapes and thus can be fitted to the ISI histograms obtained experimentally for various types of neurons. The stochastic IF neuron, however, has a definite deficiency as will be stated below.

The mean and the variance of the ISI for the positive bias μ are obtained as follows (Chhikara and Folks 1988):

$$\mathrm{E}[T_\theta] = \frac{\theta}{\mu}, \quad \mathrm{Var}[T_\theta] = \frac{\theta\sigma^2}{\mu^3}, \quad \mu > 0.$$

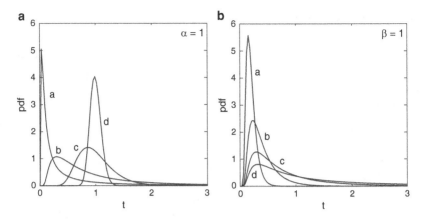

Fig. 3.34 ISI densities of the IF neuron. The function $g(t)$ of the IGD (3.23) are plotted for various values of parameters: (**a**) $\alpha = 1$, a $\beta = 0.1$; b $\beta = 1$; c $\beta = 10$; d $\beta = 100$, (**b**) $\beta = 1$, a $\alpha = 0.2$; b $\alpha = 0.4$; c $\alpha = 0.8$; d $\alpha = 1.6$, where we have set $\alpha = \theta/\mu$ and $\beta = \theta^2/\sigma^2$. (The IGD is the distribution family with the parameters α and β)

The mean ISI is the same as the ISI of the IF neuron without any random inputs ($\sigma = 0$). Noise intensity σ does not affect the mean ISI. The coefficient of variation (CV) is

$$\mathrm{CV}[T_\theta] = \frac{\sigma}{\sqrt{\mu\theta}}, \quad \mu > 0 \tag{3.24}$$

which shows that CV can take the value from 0 to infinity depending on the model parameters and that it is proportional to the noise intensity σ and also to the reciprocal of the square root of the input bias μ. By choosing the value of σ appropriately, the IF neurons can fire with a variety of irregularity holding the value of mean ISI. However, when μ is very small, which is the case that a neuron receives almost similar amount of excitatory and inhibitory inputs, both the mean and the CV of ISI become extremely big values. Therefore, the IF neuron with a large mean ISI does not show any exponential behaviour (i.e. $\mathrm{CV} \approx 1$) which is often observed in the spontaneous activities of single neurons.

3.4.2 Stochastic Leaky Integrate-and-Fire Neurons

The membrane potential of the LIF neuron (3.6) stimulated by a Gaussian white noise is described by the *Ornstein–Uhlenbeck* (OU) process:

$$dV(t) = (-V/\tau + \mu)dt + \sigma dW(t), \quad V(0) = 0 < \theta. \tag{3.25}$$

Differently from the IF neuron, the first-passage through θ is a sure event (i.e. an event with probability one) for any values of μ and $\sigma > 0$. The LIF neuron will

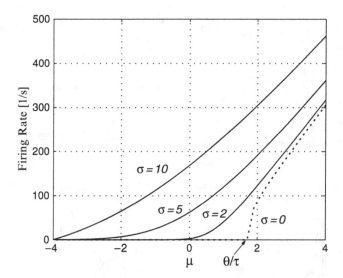

Fig. 3.35 Firing rates of the LIF neuron. *Dotted curve* shows the firing rate of the LIF neuron without a random input ($\sigma = 0$). *Solid curves* show the mean firing rate of the LIF neuron with a Gaussian white noise for various noise intensity. For the evaluation of these quantities, we have used the formula of Ricciardi and Sato (1988)

fire even when inhibitory inputs are superior to excitatory inputs. Figure 3.35 shows the firing rate of the stochastic LIF neuron as a function of the constant bias μ for various values of noise intensity σ. We can clearly see that the random inputs ensure firings of the LIF neuron for any value of μ and accelerate them. It is also shown that the firing rate as a function of μ becomes linearized as the noise intensity increases, see Clay (1976) and Yu and Lewis (1989) for the *linearization by noise*.

Figure 3.36 shows the ISI densities for the LIF neurons with various input parameters. We can see that the ISI densities cover typical features of experimentally obtained ISI histograms of single neurons such as the gamma and the exponential distributions. In particular, the densities of the LIF neurons can show a density shape with a steep rising and a high peak and also with an exponentially long tail which cannot be obtained by the ISI densities (IGD) of the IF neurons.

Figure 3.37 shows the mean and the CV of ISI as a function of the noise intensity σ with various values of μ. We can see that the values of the mean ISI for a positive bias ($\mu = 2$) are too small as a neuron model (Fig. 3.37a). When $\mu = 2$, the attractor of the deterministic part of the dynamics is above the threshold θ and therefore passage to the threshold is essentially driven by deterministic forces, where the model neuron can fire without fluctuation. On the other hand, we notice that negative biases can produce ISIs in the physiologically plausible range when noise intensity σ is enough large (see also Fig. 3.35). It tells us that the LIF neuron produces a spike train with an adequate firing rate when the fluctuation is dominant for the firing than the deterministic force by the term of μ.

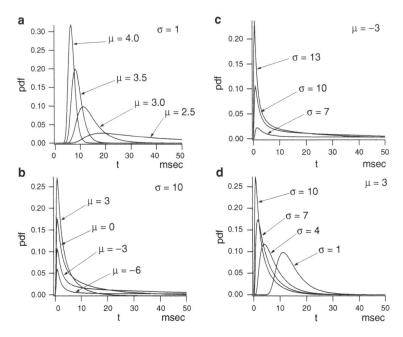

Fig. 3.36 Examples of ISI densities (histograms) of the stochastic LIF neurons. We have set $\theta = 15$ mV and $\tau = 5$ ms. They have been obtained numerically by means of the algorithm proposed in Buonocore et al. (1987)

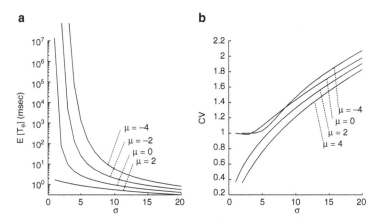

Fig. 3.37 (a) The mean $E[T_\theta]$ and (b) CV of ISIs for the OU model as a function of the noise intensity σ for various values of the parameter μ. We have set $\theta = 10$ mV and $\mu = -4, 0, 2, 4$ mV ms^{-1}. For the evaluation of these quantities, we have used the formula of Ricciardi and Sato (1988)

Figure 3.37b shows that the value of CV approaches zero and the LIF neuron acts as a pacemaker when $\sigma \to 0$ for the cases of $\mu = 2, 4$. It is also shown that the value

Fig. 3.38 Plots of the pairs of the mean $E[T_\theta]$ and CV of ISIs obtained from the LIF neuron. The firing threshold is $\theta = 10$ mV. The input parameters have been changed in the physiologically reasonable parameter ranges and the parameters which give $E[T_\theta]/\tau < 1$ are excluded. The parameters are changed in the following range: $1 \leq \sigma \leq 30$ mV ms$^{-1/2}$, $-10 \leq \mu \leq 10$ mV ms^{-1}

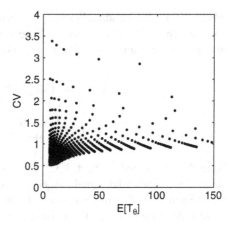

of CV of the LIF neuron ranges from 0 to ∞ mainly in accordance with the input variability σ. Low CVs are attained only when the bias μ is positively large enough to reach the threshold and the value of σ is small. For neurons which are not able to fire without random noise ($\mu = 0, -4$ in the figure), the values of CV are mostly above unity.

Figure 3.38 plots the pairs of the mean $E[T_\theta]$ and the CV of the LIF neuron for various parameter values. The parameter values have been changed in the physiologically reasonable parameter ranges and the parameters which give the mean ISI less than the membrane time constant τ are excluded. The ranges of CV for the LIF neuron is $0.5158 \leq$ CV ≤ 3.381. High CVs with a high firing rate, namely with a short mean ISI, as well as the exponential behaviour (CV ≈ 1) with a low firing rate are observed. The lower limit of CV ≈ 0.5 coincides with the result of Softky and Koch (1993) which reported that the CVs of the ISIs for neocortical neurons of awake behaving monkey ranged between 0.5 and 1.0. It is shown that the extremely high CVs are much decreased by introducing a lower boundary of the membrane potential or by introducing reversal potentials (Inoue and Doi 2007).

3.5 Stochastic Phase-Lockings and Bifurcations in Noisy Neuron Model

Next, we consider the effect of both periodic and stochastic forcing on the response characteristics of a neuronal model. Let us study the noise effects on the forced BVP model (3.17) and thus we consider a stochastic differential equation:

$$\epsilon \dot{x} = y - x + 5/6\{|x + 1| - |x - 1|\}, \tag{3.26a}$$

$$\dot{y} = -x + a + A \sin\{2\pi(t/T + \theta_0)\} + \sigma \frac{dW(t)}{dt}, \tag{3.26b}$$

where $W(t)$ is the standard Wiener process or the Brownian motion (Gardiner 1983) and the term $\sigma dW(t)/dt$ denotes a *Gaussian white noise* with noise intensity σ. We also suppose the singular limit $\epsilon = 0$ and then this equation is reduced to

$$\dot{x} = -x + a + A \sin\{2\pi(t/T + \theta_0)\} + \sigma \frac{dW(t)}{dt} , \qquad (3.27a)$$

$$y = \begin{cases} x - 5/3 & (x \geq 1), \\ x + 5/3 & (x \leq -1). \end{cases} \qquad (3.27b)$$

Note that, as stated in Sect. 3.3.2, this stochastic (singular) BVP neuron model forced by a sinusoidal signal can be considered as a combination of two stochastic LIF neuron (3.6) when a sinusoidal input is applied to.

In order to investigate the noise effects in detail, the parameter a of the BVP model (3.26) or (3.27) is set as $a = 0.0$ in the following numerical examples. Thus we are specifically considering the noise effects on the forced (piecewise-linearized) van der Pol oscillator. All the following analysis frameworks, however, do not depend on this specific choice of the parameter value a and are valid for all parameter values a, even for the case $|a| > 1$ (a non-oscillatory neuron).

3.5.1 Stochastic Phase Lockings

In the presence of noise, both variables θ_0 and θ_1 defined in (3.19) fluctuate by noise and are thus random variables Θ_0, Θ_1. We extend the deterministic map $p(\theta)$ to the case with noise as follows. Define a kernel function $g(\theta_0, \theta_1)$ using conditional probability density functions:

$$g(\theta_0, \theta_1)d\theta_1 = \Pr\{\theta_1 \leq \Theta_1 \leq \theta_1 + d\theta_1 \mid \Theta_0 = \theta_0\}. \qquad (3.28)$$

The function $g(\theta_0, \theta_1)$ can be calculated numerically without simulations of the stochastic differential equations (3.26) or (3.27) (Tateno et al. 1995; Doi et al. 1999).

Figure 3.39 is an example of the kernel function g which corresponds to the deterministic map (PTC) of Fig. 3.32b. The function g takes relatively high values along the curves of the map $p(\theta)$. Thus g can be considered as the stochastic extension of the deterministic PTC. The heights and widths of the peaks of g depend on the values of (θ_0, θ_1) and thus we can see that the effects of blurring of the deterministic map by noise are not uniform.

Using the kernel function g, we extend the system (3.20) to the noisy case as follows. Let S denote a unit circle $[0,1)$ and D the set of absolutely integrable functions with a unit L^1 norm on S. An operator P on D is defined by

$$Ph\ (\theta) = \int_S g(\theta_0, \theta)h(\theta_0)d\theta_0, \quad h \in D. \qquad (3.29)$$

Fig. 3.39 Stochastic kernel function $g(\theta_0, \theta_1)$. $a = 0.0$, $A = 0.3$, $T = 1.5$, $\sigma = 0.03$. This figure corresponds to the deterministic return map of Fig. 3.32b

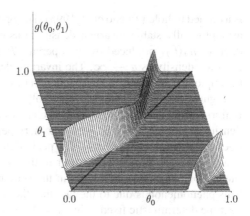

This operator is a Markov operator since it transforms a probability density function to an another probability density function (Lasota and Mackey 1994). Let $h_0(\theta) \in \mathcal{D}$ denote the probability density function of the initial phase Θ_0 when an orbit starts at the point A of the x–y phase plane of Fig. 3.30. Then the density function of Θ_1 when the orbit returns to the point A again is obtained by $h_1(\theta) = \mathcal{P}h_0(\theta)$. Thus, the deterministic mapping (3.20) is extended to the system

$$h_{n+1}(\theta) = \mathcal{P}h_n(\theta), \quad n = 0, 1, 2, \ldots. \tag{3.30}$$

Thus we investigate the asymptotic behavior of the sequence $\{h_n(\theta)\}$ of probability density functions rather than the sequence $\{\theta_n\}$ of phases. The operator $\mathcal{P}$ is called a *stochastic phase transition operator* since it is a direct extension of the deterministic PTC to the stochastic (noisy) case.

Let us list several preliminary definitions (Lasota and Mackey 1994). A function $h^*(\theta)$ is called the *invariant density* function of an operator $\mathcal{P}$ if the relation $\mathcal{P}h^* = h^*$ holds. The invariant density is *asymptotically stable* if for any initial density function $h_0 \in \mathcal{D}$

$$\lim_{n \to \infty} ||\mathcal{P}^n h_0 - h^*|| = 0,$$

where $|| \cdot ||$ is the L^1 norm on $\mathcal{S}$.

As is easily seen from its definition, the function g has the property

$$g(\theta_0, \theta_1) \geq 0, \qquad \int_{\mathcal{S}} g(\theta_0, \theta_1) d\theta_1 = 1$$

and is called a *stochastic kernel*. The inequality

$$\int_{\mathcal{S}} \inf_{\theta_0} g(\theta_0, \theta_1) d\theta_1 > 0 \tag{3.31}$$

is assumed to hold (Tateno et al. 1995). The operator with this property has a unique asymptotically stable invariant density (Lasota and Mackey 1994). Thus the sequence $\{h_n(\theta)\}$ produced by the operator $\mathcal{P}$ asymptotically approaches a unique invariant density as $n \to \infty$. The invariant density is considered as the stationary (asymptotic) probability density function of the phase Θ_n.

Figure 3.40a is an example of the evolution of the density sequence $\{h_n(\theta)\}$. The uniform initial density function $h_0(\theta)$ changes its shape and approaches an invariant density function h^* ($\approx h_{30}$) with one sharp peak, which shows a 1:1 phase locking does occur in a stochastic sense. The peak of the density functions $\{h_n(\theta)\}$ is highest near $n = 7$ and is relatively lower in the invariant density h^*, which means that fluctuations of phases are bigger in the stationary state than in the transient state. This phenomenon is due to the fact that the peaks of the kernel function $g(\theta_0, \theta_1)$ near the deterministic fixed point $\theta_0 = \theta_1 = \theta^*$ are relatively lower than the other (cf. Figs. 3.39 and 3.32b).

Figure 3.40b is also an example of the density sequence in the case of stochastic 5:3 phase locking. The initial density function $h_0(\theta)$ has a high peak near $\theta = 0.42$

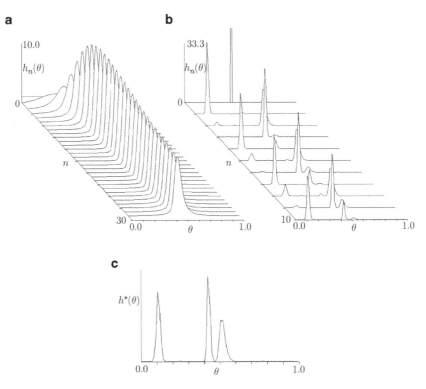

Fig. 3.40 Density evolution (sequence) $\{h_n(\theta)\}$. (**a**) Stochastic 1:1 phase locking. $A = 0.3$, $T = 1.5$, $\sigma = 0.03$. A uniform initial density function evolves to an invariant density function $h^* \approx h_{30}$. (**b**) Stochastic 5:3 phase locking. $A = 2.0$, $T = 0.85$, $\sigma = 0.02$. (**c**) Invariant density $h^* \approx h_{200}$ of (b)

which is one of the three phases to which oscillator is locked in the noise-free case. The functions h_1 and h_2 have a high mode near the other two phases of the three phases. This sequence seems to initially vary with period 3, but finally converges to an invariant density function (panel (c)) with three sharp peaks, which shows a 5:3 locking occurs in a stochastic sense.

In the noise-free case, the amplitude A of the sinusoid should be sufficiently large in order to define the deterministic map $p(\theta)$ in the case of $|a| > 1$ (non-oscillatory neuron), since the neuron may not excite (and thus orbits of the BVP model may not visit the point A of the $x-y$ phase plane) otherwise. Note that this condition is not necessary in the noisy case since there are always positive possibilities (probabilities), for all values of a, that a non-oscillatory neuron excites owing to noise.

3.5.2 Stochastic Bifurcation Diagrams

Patterns of deterministic phase lockings depend on both the amplitude A and the period T of the input. Figure 3.41 shows a deterministic bifurcation diagram of the (PTC) map $p(\theta)$ for the noise-free ($\sigma = 0$) case with a bifurcation parameter A rather than T (cf. Fig. 3.33). In this case, the period of the sinusoidal input is about a half of the intrinsic period of the van der Pol oscillator. So, for a wide range of the amplitude A, 2:1 phase lockings occur (two cycles of the input synchronize with one cycle of the oscillator). In particular, two different patterns of 2:1 phase lockings coexist; the pattern depends on both the initial phase of the input and the initial state

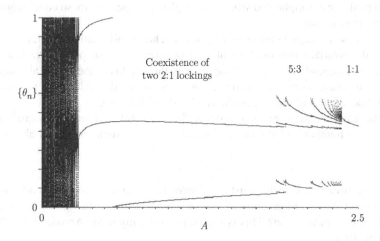

Fig. 3.41 Deterministic bifurcation diagram of the return map $p(\theta)$. The bifurcation parameter is the amplitude A rather than the period T, which is fixed to $T = 0.85$. A sequence $\{\theta_n\}, n = 101, \ldots, 600$ produced by (3.20) were plotted for each of 600 equally spaced A values on the interval $[0, 2.5]$

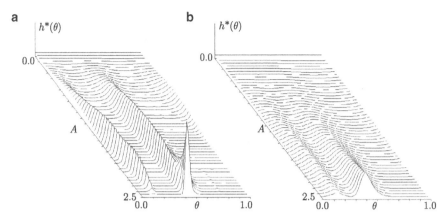

Fig. 3.42 Stochastic (invariant density) bifurcation diagram. $T = 0.85$. Invariant densities $h^*(\theta)$ were plotted for each of 50 equally spaced A values on the interval $[0, 2.5]$. (**a**) $\sigma = 0.05$ (**b**) $\sigma = 0.2$

of the oscillator. Various phase-locking patterns are bifurcated as the value of A varies. If A is large enough ($A > 2.4$), one cycle of the input can synchronize with the one cycle of the oscillator although the intrinsic periods differ by twice.

Figure 3.42 is a stochastic or a density bifurcation diagram which corresponds to the deterministic one (Fig. 3.41). Invariant density functions are plotted with various value of A for two different noise intensities: (a) $\sigma = 0.05$ and (b) $\sigma = 0.2$. In the panel (a), corresponding to the deterministic bifurcation diagram, the modes (peaks) and the shape of the invariant density are changed as the bifurcation parameter A varies. Particularly, in the range of $1.8 < A < 2.4$, the invariant densities have several peaks and complicated shapes although the shapes are not so complicated as the deterministic one.

In the case of large noise (panel b), the stochastic bifurcation diagram is very simple; the invariant densities have at most two peaks. The shapes of invariant densities do not depend on the amplitude A; only the heights of the two peaks depend on A. Thus noise, as is easily expected, washes out the dependency of the density shapes (the number of modes) on the amplitude of the input.

In the deterministic dynamical systems, the word "bifurcation" means qualitative change (the number and stability) of solutions of a system. As stated above, the equation

$$\mathcal{P}h^* = h^*$$

which the invariant density satisfies, always has a unique asymptotically stable solution and thus no bifurcation of this equation occurs in such a sense. So what is a stochastic bifurcation? This is a newly developing field (Arnold 1995, 1998; Doi et al. 1998).

Chapter 4
Whole System Analysis of Hodgkin–Huxley Systems

In Chap. 2, we have explored the dynamics of the original Hodgkin–Huxley equations of a squid giant axon where only the parameter I_{ext} was changed. The HH equations, however, include various constants or parameters whose values were determined based on physiological experiments, and thus the values inherently possess a certain ambiguity. Also, the "constants" are not really constant but change temporally. Thus, in this chapter, we study the effects of the change of the constants or parameters on the dynamics of the HH equations and consider the robustness and sensitivity of the equations; we study the bifurcation structure of the HH equations by changing their various parameters. To do so, in this chapter, we consider a slight modification of the original HH equations since the modification is mathematically more tractable.

4.1 Changing the Parameters: Sensitivity and Robustness

Consider a slight simplification of the HH equations (2.3):

$$C\frac{dv}{dt} = G(v, m, n, h) + I_{ext}, \tag{4.1a}$$

$$\frac{dm}{dt} = \frac{1}{\tau_m(v)}(m^\infty(v) - m), \tag{4.1b}$$

$$\frac{dn}{dt} = \frac{1}{\tau_n(v)}(n^\infty(v) - n), \tag{4.1c}$$

$$\frac{dh}{dt} = \frac{1}{\tau_h(v)}(h^\infty(v) - h), \tag{4.1d}$$

where the nonlinear functions $\tau_x(v)$ and $x^\infty(v)$, $(x = m, n, h)$ are denoted as follows:

$$m^\infty(v) = \frac{1}{(1 + \exp[-\bar{s}_m(v - \bar{v}_{m,1})])}, \quad \tau_m(v) = \frac{\bar{t}_m}{\cosh\left[\frac{\bar{s}_m(v - \bar{v}_{m,2})}{2}\right]}, \tag{4.2a}$$

S. Doi et al., *Computational Electrophysiology*,
DOI 10.1007/978-4-431-53862-2_4, © Springer 2010

$$n^\infty(v) = \frac{1}{(1+\exp[-\bar{s}_n(v-\bar{v}_{n,1})])}, \quad \tau_n(v) = \frac{\bar{t}_n}{\cosh\left[\dfrac{\bar{s}_n(v-\bar{v}_{n,2})}{2}\right]}, \quad (4.2b)$$

$$h^\infty(v) = \frac{1}{(1+\exp[-\bar{s}_h(v-\bar{v}_{h,1})])}, \quad \tau_h(v) = \frac{\bar{t}_h}{\cosh\left[\dfrac{\bar{s}_h(v-\bar{v}_{h,2})}{2}\right]}. \quad (4.2c)$$

Note that, differently from the original HH equations (2.3), all functions $x^\infty(v)$ (or all $\tau_x(v)$), $x = m, n, h$ have the same functional form (only the constants $\bar{s}_x$, $\bar{t}_x, \bar{v}_{x,1}, \bar{v}_{x,2}, x = m, n, h$, are different). Figure 4.1 shows the functions $x^\infty(v)$ and $\tau_x(v)$ ($x = m, n, h$) of the simplified HH equations (4.1) in solid curves and those of the original HH equations (2.3) in broken curves. First, note that, as seen from (4.2), the functions in the simplified HH equations are symmetric with respect to $v = \bar{v}_{x,i}$ ($x = m, n, h, i = 1, 2$), although the original HH equations do not have such symmetry. Secondly, note that the functions $\tau_h(v)$ and $\tau_n(v)$ in (4.2), where the values of the parameters $\bar{s}_x, \bar{t}_x, \bar{v}_{x,2}$ ($x = n, h$) are given in the caption of Fig. 4.1, are much different from those of the original HH equations.

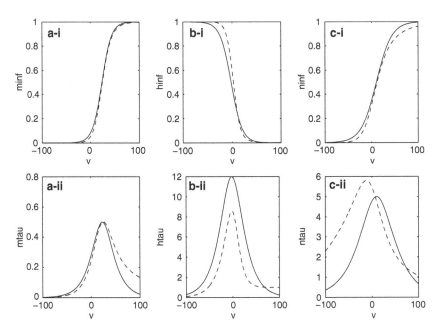

Fig. 4.1 Comparison of the functions (**a-i**) $m^\infty(v)$, (**b-i**) $h^\infty(v)$, (**c-i**) $n^\infty(v)$, (**a-ii**) $\tau_m(v)$, (**b-ii**) $\tau_h(v)$, (**c-ii**) $\tau_n(v)$ of the slight simplified HH equations (4.1) (*solid curves*) with those of the original HH equations (2.3) (*broken curves*). $\bar{s}_m = 0.1$, $\bar{v}_{m,1} = \bar{v}_{m,2} = 24.0$, $\bar{t}_m = 0.5$, $\bar{s}_h = -0.09, \bar{v}_{h,1} = \bar{v}_{h,2} = -2.0, \bar{t}_h = 12.0, \bar{s}_n = 0.06, \bar{v}_{n,1} = \bar{v}_{n,2} = 10.0, \bar{t}_n = 5.0$

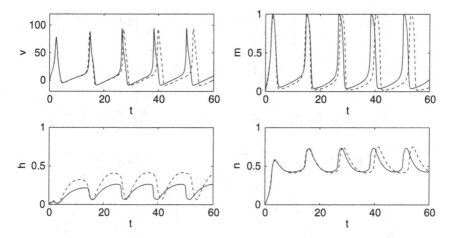

Fig. 4.2 Comparison of the solution of the slight simplified HH equations (4.1) (*solid curves*) with that of the original HH equations (2.3) (*broken curves*). $I_{ext} = 15$. $\bar{s}_m = 0.1$, $\bar{v}_{m,1} = \bar{v}_{m,2} = 24.0$, $\bar{t}_m = 0.5$, $\bar{s}_h = -0.09$, $\bar{v}_{h,1} = \bar{v}_{h,2} = -2.0$, $\bar{t}_h = 12.0$, $\bar{s}_n = 0.06$, $\bar{v}_{n,1} = \bar{v}_{n,2} = 10.0$, $\bar{t}_n = 5.0$

Different Models Show a Similar Behavior

Figure 4.2 shows the comparison of the solutions between (4.1) and (2.3). Panels (a)–(d) compare the waveforms of $v(t)$, $m(t)$, $h(t)$, $n(t)$, respectively. In spite of the big differences of the functions $\tau_h(v)$ and $\tau_n(v)$ shown in Fig. 4.1, the waveforms of $v(t)$ are very similar each other, although the periods are different slightly. (Note that a constant current is injected [$I_{ext} = 15$], and thus there is a periodic [repetitive] firing.) The waveform of $h(t)$ in the two cases are much different while the waveforms of $m(t)$ and $n(t)$ are very similar. Anyway, can we say that such neuronal systems described by (2.3) or (4.1) are robust (i.e. small change of functions or parameters do not affect the total behavior of the system)? Answer is "No" as shown in the following.

Slight Difference Leads a Big Difference of Solutions

Figure 4.3 illustrates the effect of the change of the parameter $\bar{v}_{n,1}$ of (4.1) when $I_{ext} = 46.87$. Solid and broken curves show the cases $\bar{v}_{n,1} = 5.0$ and $\bar{v}_{n,2} = 10$, respectively. Other parameters are the same as those of Fig. 4.2. As seen from panel (c-i), this change of $\bar{v}_{n,1}$ leads a very small change in the function $n^\infty(v)$ (other functions are completely the same in the two cases and overlapped, and thus broken curves are not seen). However, the solutions show completely different behavior. The period of periodic firing in solid curve is *hundred times longer* than that in broken curve (note the scale difference of abscissae in panels d and e).

As we have seen in Figs. 4.1 and 4.2, in spite of the difference of functional forms, the membrane-potential waveforms were very similar in the original HH

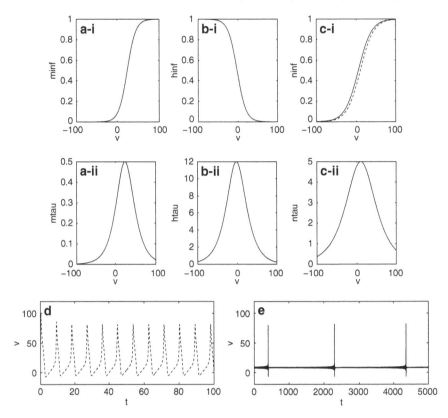

Fig. 4.3 Comparison of the functional forms and the solution of (4.1) in two cases: $\bar{v}_{n,1} = 5$ (*solid curves*) and $\bar{v}_{n,1} = 10$ (*broken curves*). Functions (**a-i**) $m^\infty(v)$, (**b-i**) $h^\infty(v)$, (**c-i**) $n^\infty(v)$, (**a-ii**) $\tau_m(v)$, (**b-ii**) $\tau_h(v)$, (**c-ii**) $\tau_n(v)$, and (**d, e**) membrane potential waveforms. $I_{\text{ext}} = 46.87$. Other parameters are the same in both cases: $\bar{s}_m = 0.1$, $\bar{v}_{m,1} = \bar{v}_{m,2} = 24.0$, $\bar{t}_m = 0.5$, $\bar{s}_h = -0.09$, $\bar{v}_{h,1} = \bar{v}_{h,2} = -2.0$, $\bar{t}_h = 12.0$, $\bar{s}_n = 0.06$, $\bar{v}_{n,2} = 10.0$, $\bar{t}_n = 5.0$

equations (2.3) and their modification (4.1). Next, we choose the parameter values so that the functions of the two equations become closer each other. Similarly to Figs. 4.1 and 4.2, Fig. 4.4 compares the two models: the original HH equations (2.3) and the simplification (4.1). (Note that Fig. 4.3 compared two solutions of the simplification (4.1) only.) Now the two models are very close each other; functional forms in both broken and solid curves are very similar. However, the behaviors of solutions are completely different. The membrane potential of the original HH equations (broken curve) fires repeatedly, while that of the simplification shows no firing after the initial firing. What has happened? We note the slight difference in the functional forms of $n^\infty(v)$ between solid and broken curves. The variable n is the activation variable of outward potassium current. Thus, we guess that the slight enhancement of the activation in low voltage region (solid curve) induces a fast outward current flow and thus inhibits a repetitive firing.

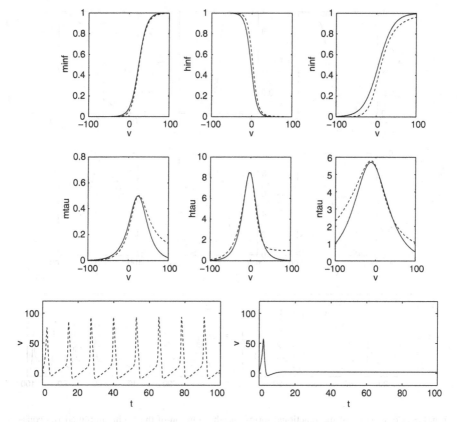

Fig. 4.4 Comparison of the functional forms and the solution of the slight simplified HH equations (4.1) (*solid curves*) with that of the original HH equations (2.3) (*broken curves*). $I_{ext} = 15$. $\bar{s}_m = 0.1, \bar{v}_{m,1} = \bar{v}_{m,2} = 24.0, \bar{t}_m = 0.5, \bar{s}_h = -0.13, \bar{v}_{h,1} = \bar{v}_{h,2} = -2.0, \bar{t}_h = 8.5, \bar{s}_n = 0.055,$ $\bar{v}_{n,1} = 5.0, \bar{v}_{n,2} = -12.0, \bar{t}_n = 5.7$

Figure 4.5 shows the similar figure to Fig. 4.4, but the value of the parameter $\bar{v}_{n,1}$ of the simplified HH equations (4.1) has been increased from 5.0 to 10.0. Now both broken and solid graphs of $n^{\infty}(v)$ are overlapped more than Fig. 4.4. Repetitive firing has been recovered! However, we note that the amplitudes of two membrane potential waveforms are different much. (Compare with the resemblance of the waveforms in Fig. 4.2.)

Through these examples, we have demonstrated:

The closeness of functional forms and/or parameter values of nonlinear dynamical systems does not mean the closeness of the solutions. Completely different behavior can be induced by a small change of model parameter and/or function. Conversely, apparently different systems often present a very similar behavior.

Therefore, we should pay much attention to the sensitivity of system's behavior on such parameter values and functional forms in model-based researches.

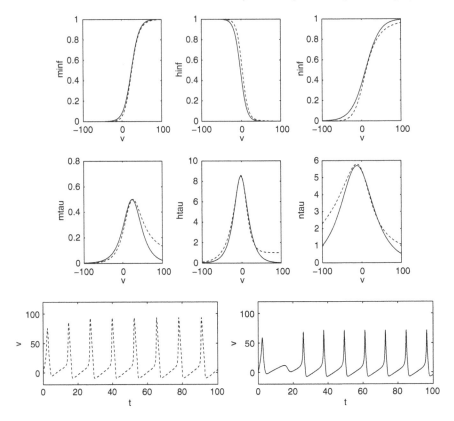

Fig. 4.5 Comparison of the functional forms and the solution of the slight simplified HH equations (4.1) (*solid curves*) with that of the original HH equations (2.3) (*broken curves*). $I_{\text{ext}} = 15$. The value of the only parameter $\bar{v}_{n,1}$ has been increased from 5.0 to 10.0 and other parameter values are the same as those of Fig. 4.4

The analysis of the dependence of system's dynamics on parameters is the bifurcation analysis. In the following, we do such bifurcation analyses in detail, in order to clarify the "whole" behavior of neuronal systems.

Exercise 4.1. Using MATLAB, try the following exercises:

- Figure 4.6 is a MATLAB script (commands) file to compare the forms of functions $x^{\infty}(v)$ and $\tau_x(v)$ ($x = m, n, h$) between the original HH equations (2.3) and its modification (4.1), as shown in Fig. 4.1. Try to run this script file. Try also to run it with changing several parameters.
- Figure 4.7 is also a MATLAB script file to compare the solutions between the original HH equations (2.3) and its modification (4.1), as shown in Fig. 4.2. Try to execute this script file (with the two script files shown in Figs. 4.8 and 4.9) on MATLAB. Try also to run it with changing several parameters and confirm the results described in this section (Figs. 4.3, 4.4 and 4.5). ∎

```
%HHtypeFncs_1.m
clear all;
v=[-100+0.001:0.5:100];
am = 0.1*(v-25)./(1-exp(-(v-25)/10));
bm = 4*exp(-v/18);
ah = 0.07*exp(-v/20);
bh = 1./(1+exp(-(v-30)/10));
an = 0.01*(v-10)./(1-exp(-(v-10)/10));
bn = 0.125*exp(-v/80);
minf = am./(am+bm); mtau = 1./(am+bm);
hinf = ah./(ah+bh); htau = 1./(ah+bh);
ninf = an./(an+bn); ntau = 1./(an+bn);
%
Sm=0.1; Vm=24.0; Tm=0.5;
Sh=-0.13; Vh=-2.0; Th=8.5;
Sn=0.055; Vn1=10.0; Vn2=-12.0; Tn=5.7;
Minf = 1./(1+exp(-Sm*(v-Vm))); Mtau = Tm./cosh(Sm*(v-Vm)/2);
Hinf = 1./(1+exp(-Sh*(v-Vh))); Htau = Th./cosh(Sh*(v-Vh)/2);
Ninf = 1./(1+exp(-Sn*(v-Vn1))); Ntau = Tn./cosh(Sn*(v-Vn2)/2);
%
subplot(2,3,1); plot(v,minf,'-b',v,Minf,'-r'); ylabel('minf');
xlabel('v');
subplot(2,3,4); plot(v,mtau,'-b',v,Mtau,'-r'); ylabel('mtau');
subplot(2,3,2); plot(v,hinf,'-b',v,Hinf,'-r'); ylabel('hinf');
subplot(2,3,5); plot(v,htau,'-b',v,Htau,'-r'); ylabel('htau');
subplot(2,3,3); plot(v,ninf,'-b',v,Ninf,'-r'); ylabel('ninf');
subplot(2,3,6); plot(v,ntau,'-b',v,Ntau,'-r'); ylabel('ntau');
```

Fig. 4.6 MATLAB script file: HHtypeFncs_1.m which compares the functional forms between original HH equations (2.3) and its modification (4.1)

4.2 Bifurcations of the Hodgkin–Huxley Neurons

In the rest of this chapter, we consider the slightly simplified HH equations (4.1) only (we simply call these equations as the HH equations or the HH model unless any confusions occur) and consider the sensitivity of the HH model on the parameter changes.

Neurons receive inputs from other neurons. Thus, the external current I_{ext} of (4.1) is the most important parameter. Figure 4.10a shows the bifurcation diagram of the HH equations (4.1) similar to the diagram (Fig. 2.15) of the original HH equations (2.3). The abscissa denotes the bifurcation parameter I_{ext} and the ordinate denotes the membrane potential v where the maximum value of v is plotted for a periodic (oscillatory) solution. Solid and dotted curves denote stable and unstable equilibrium points, respectively. The filled (open) circles denote stable (unstable, resp.) periodic solutions. The parameter values of the HH equations (4.1) in Fig. 4.10 are the same as those of Fig. 4.2.

```
%HHtypeSimu_1.m
clear all;
global Iext TTm TTh TTn
global Gna Gk Gl Vna Vk Vl
global sm vm tm sh vh th sn vn tn
global Sm Vm Tm Sh Vh Th Sn Vn1 Vn2 Tn
Gna = 120; Gk = 36; Gl = 0.3;
Vna = 115; Vk = -12;    Vl = 10.599;
Sm=0.1; Vm=24.0; Tm=0.5;
Sh=-0.13; Vh=-2.0; Th=8.5;
Sn=0.055; Vn1=10.0; Vn2=-12.0; Tn=5.7;
Iext=15; TTm=1; TTh=1; TTn=1;
TIME1=100; TIME2=TIME1;
[t1,x]= ode23('HH_1',[0 TIME1],[0, 0, 0, 0]);
[t2,y]= ode23('HHtype2',[0 TIME1],[0, 0, 0, 0]);
clf;
subplot(3,2,1); plot(t1,x(:,1),'-b'); xlabel('t');ylabel('v');
axis([0 TIME1 -20 120]);
subplot(3,2,3); plot(t2,y(:,1),'-r'); xlabel('t');ylabel('v');
axis([0 TIME2 -20 120]);
```

Fig. 4.7 MATLAB script file: HHtypeSimu_1.m which compares the *solutions* between original HH equations (2.3) and its modification (4.1)

```
%HH_1.m
function [xdot,xinit,option] = HH_1(t, x)
global Iext TTm TTn TTh
global Gna Gk Gl Vna Vk Vl
    C=1;
    v=x(1); m=x(2); h=x(3); n=x(4);
    am = 0.1*(v-25)/(1-exp(-(v-25)/10));
    bm = 4*exp(-v/18);
    ah = 0.07*exp(-v/20);
    bh = 1/(1+exp(-(v-30)/10));
    an = 0.01*(v-10)/(1-exp(-(v-10)/10));
    bn = 0.125*exp(-v/80);
%
xdot = [
(Gna*m^3*h*(Vna-v) + Gk*n^4*(Vk-v) + Gl*(Vl-v) + Iext)/C;
(am*(1-m)-bm*m)/TTm; (ah*(1-h)-bh*h)/TTh; (an*(1-n)-bn*n)/TTn
        ];
```

Fig. 4.8 MATLAB script file: HH_1.m which describes the original HH equations (2.3)

At the point HB1 ($I_{ext} = 1.93$), unstable periodic solutions are bifurcated from the equilibrium point by the (sub-critical or unstable) Hopf bifurcation. Stable periodic solutions are bifurcated by the (super-critical or stable) Hopf bifurcation at the point HB2 ($I_{ext} = 282.92$). At the point DC, the double-cycle bifurcation or

```
%HHtype2.m
function [xdot,xinit,option] = HHtype2(t, x)
global Iext TTm TTn TTh
global Gna Gk Gl Vna Vk Vl
global Sm Vm Tm Sh Vh Th Sn Vn1 Vn2 Tn
C=1;
v=x(1); m=x(2); h=x(3); n=x(4);
%
Minf = 1/(1+exp(-Sm*(v-Vm))); Mtau = Tm/cosh(Sm*(v-Vm)/2);
Hinf = 1/(1+exp(-Sh*(v-Vh))); Htau = Th/cosh(Sh*(v-Vh)/2);
Ninf = 1/(1+exp(-Sn*(v-Vn1))); Ntau = Tn/cosh(Sn*(v-Vn2)/2);
%
xdot = [
(Gna*m^3*h*(Vna-v) + Gk*n^4*(Vk-v) + Gl*(Vl-v) + Iext)/C;
(Minf-m)/(Mtau*TTm); (Hinf-h)/(Htau*TTh); (Ninf-n)/(Ntau*TTn)
];
```

Fig. 4.9 MATLAB script file: HHtype2.m which describes the modified HH equations (4.1)

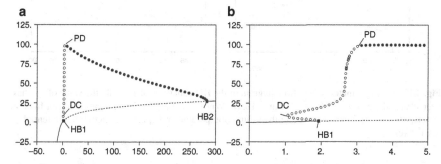

Fig. 4.10 One-parameter bifurcation diagram of the HH equations (4.1) with respect to the bifurcation parameter I_{ext}. Other parameter values are the same as those of Fig. 4.2. Hopf bifurcations occur at HB1 ($I_{ext} = 1.934$, eigenvalues: $0.000003132 \pm 0.436584i$, -0.0941944, -4.65870) and at HB2 ($I_{ext} = 282.916$, eigenvalues: $0.000000220614 \pm 0.969227i$, -0.181518, -14.7220). (**a**) Whole diagram. (**b**) Magnification of (a). In both panels, *abscissa* and *ordinate* denote the bifurcation parameter I_{ext} and the membrane potential v, respectively

saddle-node bifurcation of periodic solution occurs and a pair of *unstable* periodic solutions are generated. At the point PD, the period-doubling bifurcation occurs and the stability of periodic solution changes. Panel (b) is the magnification of (a) near HB1. Differently from the bifurcation diagram (Fig. 2.15) of the original HH equations, there is no region where both a stable equilibrium point and a periodic orbit co-exist. Despite such a difference, whole nature of both Figs. 2.15 and 4.10 are similar. Therefore, both solutions shown in Fig. 4.2 of the HH equations (4.1) and the original HH equations (2.3) much resembled each other.

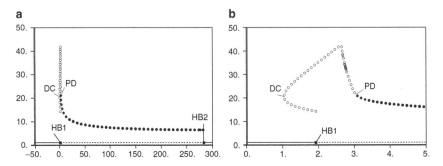

Fig. 4.11 Period of periodic solutions shown in (**a**) Fig. 4.10a and (**b**) Fig. 4.10b. In both panels, *abscissa* and *ordinate* denote I_{ext} and the period, respectively

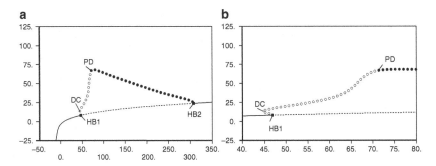

Fig. 4.12 One-parameter bifurcation diagram of the HH equations (4.1). Only the value of $\bar{v}_{n,1}$ has been changed from 10.0 to 5.0 from Fig. 4.10. (**a**) Whole diagram. (**b**) Magnification of (a). In both panels, *abscissa* and *ordinate* denote the bifurcation parameter I_{ext} and the membrane potential v, respectively

Figure 4.11 shows the period of periodic solutions shown in Fig. 4.10. We can see that, also with respect to the period of periodic solution, the HH equations (4.1) and the original HH equations (2.3) resemble each other (compare this figure with Fig. 2.16). At the Hopf bifurcation point HB1, the equilibrium point of (4.1) possesses *approximately* pure imaginary eigenvalues: $0.000003132 \pm 0.436584i$. The period of (unstable) periodic orbit which is close to HB1 is determined by the imaginary part as $2\pi/0.436584 \approx 14.39$. Please check that this period coincides with that of Fig. 4.11.

Figures 4.12 and 4.13 are the similar bifurcation diagrams to Figs. 4.10 and 4.11, where only the value of $\bar{v}_{n,1}$ has been changed from 10.0 to 5.0. In spite of this very small difference, the whole nature of both bifurcation diagrams Figs. 4.10 and 4.12 are completely different each other. Particularly, the position of HB1 has moved much in right direction, whereas that of HB2 has moved slightly. Also, the range between HB1 and PD has been expanded significantly. The amplitudes of periodic solutions (repetitive firing) have been decreased much, while the periods of periodic solutions are not so different between Figs. 4.11 and 4.13. This significant change of bifurcation structure is the reason of the unexpectedly different firing shown in

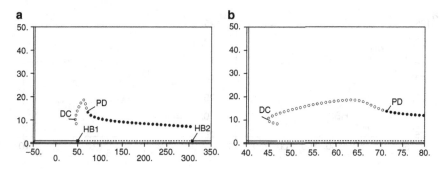

Fig. 4.13 Period of periodic solutions shown in (**a**) Fig. 4.12a and (**b**) Fig. 4.12b. In both panels, *abscissa* and *ordinate* denote I_{ext} and the period, respectively

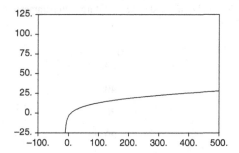

Fig. 4.14 One-parameter bifurcation diagram of the HH equations (4.1) with respect to the bifurcation parameter I_{ext}. Other parameter values are the same as those of Fig. 4.4. *Abscissa* and *ordinate* denote the bifurcation parameter I_{ext} and the membrane potential v, respectively

Fig. 4.3e. (Note that the value $I_{\text{ext}} = 46.87$ of Fig. 4.3e locates near HB1 in Fig. 4.12 where only unstable solutions exist. Strictly speaking, the solution in Fig. 4.3e is the result of more complicated bifurcations taken place near HB1.)

Next, consider the case that the parameter values of the HH equations (4.1) are chosen so that the functions $x^{\infty}(v)$ and $\tau_x(v)$ ($x = m, n, h$) of (4.1) could become close to those of the original HH equations (2.3). Figure 4.14 shows the bifurcation diagram where the parameter values are the same as those of Fig. 4.4 where no firing occurs. We note again that both HH equations of (4.1) and (2.3) resemble each other, as shown in Fig. 4.4. However, as shown in the bifurcation diagram of Fig. 4.14, the HH equations (4.1) have neither Hopf bifurcation nor firing even when we increase the value of I_{ext}, whereas the original HH equations (2.3) show rich firing phenomena.

Figures 4.15 and 4.16 respectively show the bifurcation diagram and the periods of periodic solutions when only the value of $\bar{v}_{n,1}$ is increased from 5 to 10.0. Hopf bifurcations and repetitive firing (periodic solutions) have been recovered. Note, however, that the range of the external current I_{ext} where periodic solutions exist has been shrunk much in contrast to the previous bifurcation diagrams shown in Figs. 4.10 and 4.12. Again, we would like to emphasize that the difference of the

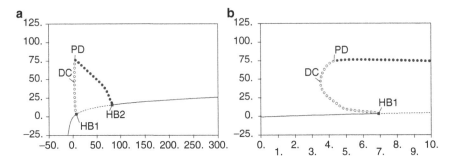

Fig. 4.15 One-parameter bifurcation diagram of the HH equations (4.1) with respect to the bifurcation parameter I_{ext}. Only the value of $\bar{v}_{n,1}$ is increased to 10.0 from Fig. 4.14: All parameter values except for I_{ext} are the same as those of Fig. 4.5. (**a**) Whole diagram. (**b**) Magnification of (a). In both panels, *abscissa* and *ordinate* denote the bifurcation parameter I_{ext} and the membrane potential v, respectively

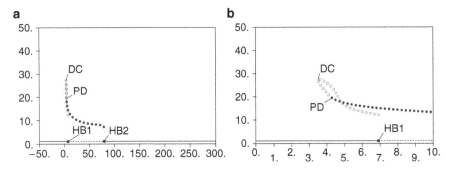

Fig. 4.16 Period of periodic solutions shown in (**a**) Fig. 4.15a and (**b**) Fig. 4.15b. In both panels, *abscissa* and *ordinate* denote I_{ext} and the period, respectively

parameter value between Figs. 4.14 and 4.15 are very small, although the total behavior of both bifurcation diagrams is completely different: The solutions of the HH equations (4.1) are very sensitive to the parameter $\bar{v}_{n,1}$.

In contrast to the above examples, next example will show us the *insensitivity (robustness)* and will make us very confusing! In the bifurcation diagram of Fig. 4.17, only the value of $\bar{v}_{n,2}$ has been increased to 40.0 from -12.0 of Fig. 4.15. Comparing Fig. 4.17 with Fig. 4.15, we can see that the whole feature of the bifurcation diagrams are very similar each other *irrespective of the big difference* of the $\bar{v}_{n,2}$ values: 40 and -12. Of course, details of the diagrams have changed: The amplitudes of periodic solutions decreased by the increase of $\bar{v}_{n,2}$. The loci of Hopf bifurcation points HB1 and HB2 moved rightward slightly. The locus of period-doubling bifurcation point PD moved rightward much. The periods of periodic solutions decreased slightly, particularly in the small I_{ext} region (Fig. 4.18). Note that the parameter $\bar{v}_{n,2}$ affects only the function $\tau_n(v)$ in (4.1) and does not affect *the position* of the equilibrium points of (4.1) since $\tau_n(v)$ does not affect the solution of $dv/dt = dm/dt = dn/dt = dh/dt = 0$ at all (please examine the differ-

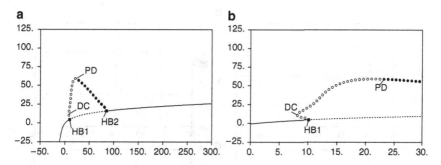

Fig. 4.17 One-parameter bifurcation diagram of the HH equations (4.1) with respect to the bifurcation parameter I_{ext}. Only the value of $\bar{v}_{n,2}$ is increased to 40.0 from Fig. 4.15. (**a**) Whole diagram. (**b**) Magnification of (**a**). In both panels, *abscissa* and *ordinate* denote the bifurcation parameter I_{ext} and the membrane potential v, respectively

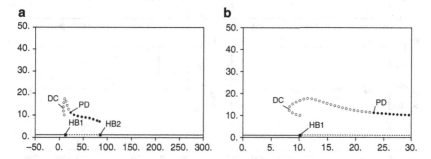

Fig. 4.18 Period of periodic solutions shown in (**a**) Fig. 4.17a and (**b**) Fig. 4.17b. In both panels, *abscissa* and *ordinate* denote I_{ext} and the period, respectively

ence of solid and broken curves between Figs. 4.17 and 4.15). However, $\tau_n(v)$ *does affect the stability* of equilibrium points. In fact, the loci of Hopf bifurcation points have been changed. Of course, $\tau_n(v)$ affects the amplitudes and periods of periodic solutions. Anyway, the big change of $\tau_n(v)$ value does not change the whole bifurcation structure much: the HH equations (4.1) are not sensitive to $\tau_n(v)$ and are robust.

So far, using several examples, we have explored and shown that the HH equations (4.1) are sensitive to some parameters and insensitive to other parameters. What makes this difference of sensitivity? The two-parameter bifurcation analysis in the following section will clarify the reason of this difference.

4.3 Two-Parameter Bifurcation Analysis of the HH Equations

Figure 4.19 shows several *two-parameter* bifurcation diagrams of the HH equations (4.1). The abscissa is the bifurcation parameter I_{ext} similarly to the previous one-parameter bifurcation diagrams, while the ordinate denotes also another

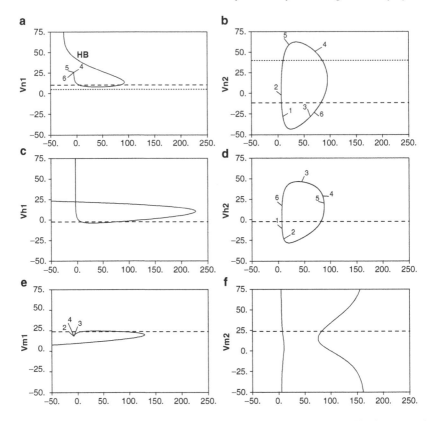

Fig. 4.19 Two-parameter bifurcation diagram of the HH equations (4.1). Abscissae are the bifurcation parameter I_{ext} for all panels. Ordinates are also bifurcation parameters: (**a**) $\bar{v}_{n,1}$ (**b**) $\bar{v}_{n,2}$ (**c**) $\bar{v}_{h,1}$ (**d**) $\bar{v}_{h,2}$ (**e**) $\bar{v}_{m,1}$ (**f**) $\bar{v}_{m,2}$. Parameter values other than two bifurcation parameters for each panel are the same as those of Fig. 4.5

bifurcation parameter. All parameters of (4.1) other than the two bifurcation parameters in each panel are the same as those of Fig. 4.15 (Fig. 4.5). In the panel (a), the ordinate is $\bar{v}_{n,1}$. The solid curve called a *bifurcation curve* denotes the loci of Hopf bifurcations; if the values of parameters $(I_{ext}, \bar{v}_{n,1})$ are chosen on this curve, a Hopf bifurcation occurs. The horizontal broken line shows the line with $\bar{v}_{n,1} = 10.0$. If we change the parameter I_{ext} along this line, the *one-parameter* bifurcation diagram shown in Fig. 4.15 is obtained. We can confirm that the I_{ext} values of two intersections of the broken line and the bifurcation curve are 6.9 and 82.0 which are exactly the same as those of the two Hopf bifurcation points (HB1 and HB2) in the one-parameter bifurcation diagram (Fig. 4.15).

The horizontal dotted line shows the line with $\bar{v}_{n,1} = 5.0$. Since this line never intersect the Hopf bifurcation curve, we cannot see any Hopf bifurcations when I_{ext} is changed along this line. The one-parameter bifurcation diagram along the dotted line is Fig. 4.14. In fact, no bifurcation occurs in Fig. 4.14. This is the reason why two bifurcation diagrams in Figs. 4.14 and 4.15 (also two solutions in Figs. 4.4 and 4.5) were much different each other, in spite of the small difference of the $\bar{v}_{n,1}$ values. The difference of the relative positions of the broken line ($\bar{v}_{n,1} = 10$) and the dotted line ($\bar{v}_{n,1} = 5$) compared to the Hopf bifurcation curve is the cause of the sensitivity of the HH equations (4.1) to the $\bar{v}_{n,1}$ value.

Figure 4.19b is a similar bifurcation diagram to panel (a), but the ordinate is the other bifurcation parameter $\bar{v}_{n,2}$. The broken line is the line with $\bar{v}_{n,2} = -12$. Since this value of $\bar{v}_{n,2}$ is the same as that of Fig. 4.14, the one-parameter bifurcation diagram along this line is also Fig. 4.14. Confirm that the loci (I_{ext} values) of the two intersections of the broken line and the solid bifurcation curve coincide with those of the Hopf bifurcations (HB1 and HB2) of Fig. 4.14 (and also coincide with those of Fig. 4.19a). The dotted line is the line with $\bar{v}_{n,2} = 40$ which is the same as that of Fig. 4.17. Thus, the one-parameter bifurcation diagram along this line is Fig. 4.17. In Fig. 4.19b, the loci of two intersections of the dotted line and the HB bifurcation curve are very similar to the loci of two intersections of the *broken* line and the HB bifurcation curve. This is the reason why the two bifurcation diagrams shown in Figs. 4.15 and 4.17 much resemble each other, in spite of the big difference of the $\bar{v}_{n,2}$ value. As seen from the way that the broken line intersects with the HB bifurcation curve, the solutions and Hopf bifurcations of (4.1) are not sensitive but are robust to the change of the parameter $\bar{v}_{n,2}$.

Other panels (c)–(f) are similar two-parameter bifurcation diagrams and their ordinates are $\bar{v}_{h,1}$, $\bar{v}_{h,2}$, $\bar{v}_{m,1}$, $\bar{v}_{m,2}$, respectively. On the broken line of each panel, the value of the ordinate (the second bifurcation parameter) is the same as that of Fig. 4.14, the loci (I_{ext} values) of intersections of the broken line and the solid bifurcation curve are the same in all panels. From these bifurcation diagrams, we can find that the HH equations (4.1) are very sensitive to both $\bar{v}_{h,1}$ and $\bar{v}_{m,1}$, but not sensitive to both $\bar{v}_{h,2}$ and $\bar{v}_{m,2}$. These results are consistent with the case of $\bar{v}_{n,1}$ and $\bar{v}_{n,2}$. Thus, the two-parameter bifurcation diagrams shown in Fig. 4.19 lead to an important implication that the HH equations (4.1) are very sensitive to the functions $x^{\infty}(v)$, ($x = m, n, h$), but not sensitive to $\tau_x(v)$, ($x = m, n, h$).

As explained in the above section, the period of (infinitesimally small) periodic orbits bifurcated from an equilibrium point through a Hopf bifurcation can be determined by the imaginary parts of complex eigenvalues at the equilibrium point. Figure 4.20 shows such periods along the Hopf bifurcation curves in Fig. 4.19. The abscissa is the bifurcation parameter I_{ext} similarly to Fig. 4.19 while the ordinate shows the period. For example, at the point labeled as 6 of Fig. 4.19a, the Hopf bifurcation curve terminates. From Fig. 4.20a, we can see that the period grow up rapidly (to infinity possibly) near the termination point (the point labeled as 6 is outside the area of Fig. 4.20a). In Fig. 4.19b, the Hopf bifurcation curve is a closed curve, while the curve in Fig. 4.20b is not. This means that the periods of periodic orbits along the Hopf bifurcation curve in Fig. 4.19b do not depend on the value of $\bar{v}_{n,2}$ but do depend on only the I_{ext} value of the abscissa.

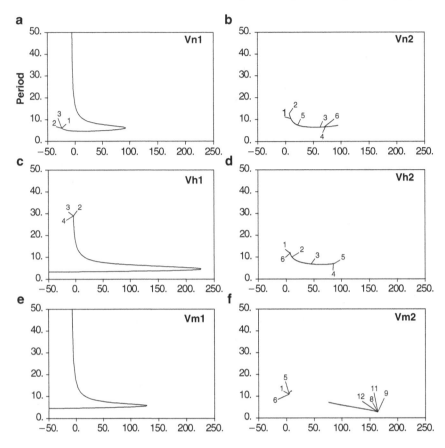

Fig. 4.20 Periods at the Hopf bifurcations of the two-parameter bifurcation diagram Fig. 4.19. *Abscissae* are the bifurcation parameter I_{ext} and *ordinates* are the period. Each panel corresponds to that of Fig. 4.19

4.4 Numerical Bifurcation Analysis by XPPAUT

In this section, we briefly explain how to execute the two-parameter bifurcation analyses shown in the above section, using the XPPAUT which was also briefly explained in Sect. 1.5. Figure 4.21 shows the `HHtype.ode` file to analyze the HH equations (4.1). In order to obtain the two-parameter bifurcation diagram of Fig. 4.19b, for example, perform the following steps:

1. Start up XPPAUT and open (or make) the `HHtype.ode` file shown in Fig. 4.21.
2. Run the commands `Initialconds` and `(G)o`, then `Initialconds` and `(L)ast` several times to obtain a solution sufficiently converged to a stable equilibrium point.
3. Use `File` and `Auto` commands to open the AUTO window.

```
# modified Hodgkin-Huxley equation HHtype.ode
init v=-0.015124 m=0.083057 h=0.45546 n=0.35414

Minf = 1/( 1+exp(-sm*(v-vm1)) )
Mtau = tm/cosh( sm*(v-vm2)/2 )
Hinf = 1/( 1+exp(-sh*(v-vh1)) )
Htau = th/cosh( sh*(v-vh2)/2 )
Ninf = 1/( 1+exp(-sn*(v-vn1)) )
Ntau = tn/cosh( sn*(v-vn2)/2 )

v' = (Gna*m^3*h*(Vna-v) + Gk*n^4*(Vk-v) + Gl*(Vl-v) + Iext)/C
m' = (Minf-m)/(Mtau)
h' = (Hinf-h)/(Htau)
n' = (Ninf-n)/(Ntau)

par Iext=0 C=1
par Gna=120, Gk=36, Gl=0.3
par Vna=115, Vk=-12, Vl=10.599

par sm=0.1, vm1=24.0,vm2=24.0, tm=0.5

#par sh=-0.09, vh1=-2.0,vh2=-2.0, th=12.0
par sh=-0.13 vh1=-2.0 vh2=-2.0 th=8.5

#par sn=0.06, vn1=10.0, vn2=10.0, tn=5.0
par sn=0.055 vn1=10.0 vn2=-12.0 tn=5.7

@ xplot=t,yplot=v
@ total=100,dt=.03,xlo=0,xhi=100,ylo=-20,yhi=120
@ meth=runge-kutta
@ bound=200
@ ntst=50 nmax=1000 npr=500 parmin=0 parmax=100 dsmax=2
done
```

Fig. 4.21 The ode file of XPPAUT: HHtype.ode

4. Use Parameter command to open the Parameters window where we change the second parameter *Par2 (which corresponds to the ordinate of Fig. 4.19b) from C to vn2.

5. Use Numerics to change several parameter values for numerical bifurcation analysis: Change Nmax (maximum number of computational steps) from 1,000 to 250 (this may be sufficient), Ds (usual step size of changing the bifurcation parameter) from 0.02 to 0.1 (this is small enough), both Par Min and Par Max (the range of changing the main (first) bifurcation parameter (I_{ext})) from "0 and 100" to "−50 and 250" (the same scale as Fig. 4.19b). Then the window will become as Fig. 4.22. Next, click on Ok.

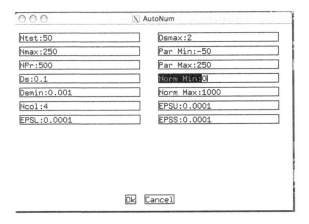

Fig. 4.22 AUTO Numerics Window

6. Use Axes and hI-lo to change the range of both abscissa and ordinate of the graph: Xmin:-50, Ymin:-20, Xmax:250, Ymax:120.
7. Run and Steady state to obtain a one-parameter bifurcation diagram of equilibrium points.
8. Use Grab and tab (or →, ← keys on a keyboard) to move the cursor on the Hopf bifurcation point (usually labeled by 2) of the bifurcation diagram. Then hit return or enter key to grab the point.
9. Use Run and Two Param to obtain a two-parameter bifurcation diagram. Finally, use Axes and Two par to change *Y-axis from V to vn2, Ymin from −20 to −50, Ymax from 120 to 75. Then, reDraw gives us the two-parameter bifurcation diagram shown in Fig. 4.23.
10. Note that, if you failed in some of the above steps, close the AUTO window and return to Step2. Then, pay attention to the fact that the values of the parameters as well as those of the variables v, m, h, n have been changed by bifurcation analysis: Reset the values to initial or desired values. Also note that when we return to the AUTO window, previous results still remain: Reset all results of the previous bifurcation analysis by File and Reset diagram (and/or Clear grab) in the AUTO window. If you still fail, return to Step1: quit XP-PAUT and restart it!

Exercise 4.2. Using the XPPAUT and the HHtype.ode file shown in Fig. 4.21, try the following exercises:

• Draw the one-parameter bifurcation diagram (bifurcation parameter is I_{ext}) such as those shown in Fig. 4.15.
• Changing minimum and maximum values of axes, magnify the one-parameter bifurcation diagram near the Hopf bifurcation.
• Return to the XPP main window, then obtain several solutions of the HH equations with changing the value of I_{ext} (draw not only "v vs. t" plot but also

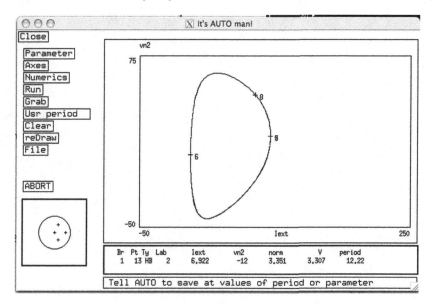

Fig. 4.23 Two-parameter bifurcation diagram made by XPPAUT similar to Fig. 4.19b

"*m* vs. *v*" plot and "*h* vs. *v*" plot, etc.). When you make these simulations, consult the (magnified) one-parameter bifurcation diagram.

- Changing the value of parameters (for example vn1), repeat above steps.
- Try to make the two-parameter bifurcation diagram of Fig. 4.23, then try to make other two-parameter bifurcation diagrams shown in Fig. 4.19. □

Chapter 5
Hodgkin–Huxley-Type Models of Cardiac Muscle Cells

Following the HH formalism introduced in Chap. 2, various kinds of HH-type models of neurons and other excitable cells are proposed (Canavier et al. 1991; Chay and Keizer 1983; Cronin 1987; Gerber and Jakobsson 1993; Hayashi and Ishizuka 1992; Keener and Sneyd 1998; Noble 1995; Rinzel 1990; Traub et al. 1991), and are analyzed (Alexander and Cai 1991; Av-Ron 1994; Bertram 1994; Bertram et al. 1995; Butera 1998; Canavier et al. 1993; Chay and Rinzel 1985; Doi and Kumagai 2005; Guckenheimer et al. 1993; Maeda et al. 1998; Rush and Rinzel 1994; Schweighofer et al. 1999; Terman 1991; Tsumoto et al. 2003, 2006; Yoshinaga et al. 1999). The HH-type equations include many variables depending on the number of different ionic currents and their gating variables considered in the equations, whereas the original HH equations possess only four variables (a membrane voltage, activation and inactivation variables of Na^+ current and an activation variable of K^+ current). Among the diverse family of HH-type equations, this chapter explores the dynamics and the bifurcation structure of the HH-type equations of heart muscle cells (cardiac myocytes).

5.1 Action Potentials in a Heart

A heart repeats contraction and relaxation, and these motions are controlled by electrical signals (action potentials). Periodic electrical signals are generated in sino-atrial node (cardiac pacemaker) and they are propagated to the whole heart. The regular generation of the periodic electrical signals and their regular propagations (conductions) are essential for the pumping function of the heart. Figure 5.1 shows various action potentials of different parts of heart. The action potential waveforms differ part by part in the heart. Therefore, the waveform itself is important for the normal function of the heart (Noble 1975).

Figure 5.2 shows a typical action potential waveform in ventricular myocardial cell with five "phases." Phase 0 is called the rapid depolarization phase. In this phase, Na^+ currents flow into the cell, and the membrane potential is increased to positive value rapidly. K^+ currents outflow from the cell in phase 1 called the overshoot and early repolarization phase. Phase 2 is kept by a balance between inward

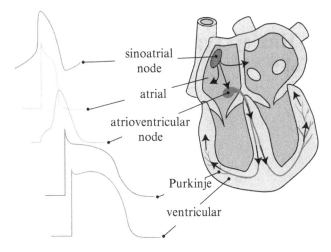

Fig. 5.1 Schematic illustrations of the heart and various action potentials of different parts of the heart

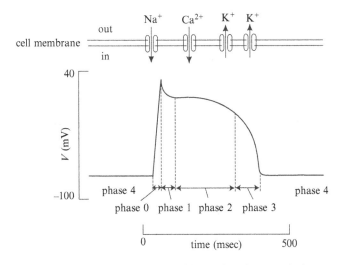

Fig. 5.2 Schematic illustration of an action potential waveform in a ventricular myocardial cell. In each phase of the action potential, specific ions move between the inside and outside of the cell membrane

Ca^{2+} current and outward K^+ current, and it is called the plateau phase. Phase 3 and phase 4 are due to outward K^+ current, and they are called the repolarization phase and the resting potential phase, respectively. The period from phase 0 to phase 3 is called a refractory period. During this period, it is difficult or impossible for the cell to respond to external stimuli.

5.2 Pacemaker Cell Model

5.2.1 The YNI Model of Cardiac Pacemaker Cell

At first, let us consider a Hodgkin–Huxley-type model of cardiac pacemaker cells (sinoatrial-node cells). The Yanagihara–Noma–Irisawa (YNI) model (Yanagihara et al.1980) of sinoatrial-node cells is a typical (and rather classical) Hodgkin–Huxley-type model of cardiac cells and is described as follows:

$$\frac{dV}{dt} = -\frac{1}{C}(I_{Na} + I_s + I_h + I_K + I_l), \tag{5.1a}$$

$$\frac{dx}{dt} = \alpha_x(V)(1 - x) - \beta_x(V)x, \quad (x = m, h, d, f, q, p), \tag{5.1b}$$

where V is the membrane potential and x ($= m, h, d, f, q, p$) are the gating variables. The YNI model is described by differential equations with seven variables. In (5.1a), there are five ionic currents: I_{Na} is the sodium current, I_s is the slow inward current, I_h is the hyperpolarization-activated current, I_K is the potassium current and I_l is the leak current. (Notice that an inward current through ionic channels is denoted as a "negative current" in the YNI model. See the minus sign of the r.h.s. of (5.1a).) Figure 5.3 illustrates these ionic currents or ionic channels. Note that, differently from the original HH model (2.3) of squid giant axon, there are five kinds of different ionic currents considered. Equations which describe the five ionic currents are as follows:

$$I_{Na} = c_{Na}G_{Na}m^3h(V - 30), \quad G_{Na} = 0.5;$$

$$I_s = c_sG_s(0.95d + 0.05)(0.95f + 0.05)\left(\exp\left(\frac{V - 30}{15}\right) - 1\right), \quad G_s = 12.5;$$

$$I_h = c_hG_hq(V + 45), \quad G_h = 0.4;$$

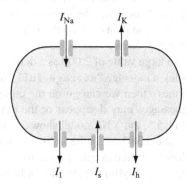

Fig. 5.3 Schematic diagram of the Yanagihara–Noma–Irisawa (YNI) model (5.1) which has five ionic channels

$$I_K = c_K G_K p \frac{\exp\left(0.0277(V + 90)\right) - 1}{\exp\left(0.0277(V + 40)\right)}, \quad G_K = 0.6;$$

$$I_1 = c_1 G_1 \left(1 - \exp\left(-\frac{V + 60}{20}\right)\right), \quad G_1 = 0.8;$$

where the parameters c_{Na}, c_s, c_h, c_K, c_1 are the coefficients which change the conductances of ionic currents.

The functions $\alpha_x(V)$ and $\beta_x(V)$ $(x = m, h, d, f, q, p)$ in (5.1b) are as follows:

$$\alpha_m(V) = \frac{V + 37}{1 - \exp\left(-\frac{V + 37}{10}\right)}, \quad \beta_m(V) = 40 \exp\left(-\frac{V + 62}{17.8}\right);$$

$$\alpha_h(V) = 1.209 \times 10^{-3} \exp\left(-\frac{V + 20}{6.534}\right), \quad \beta_h(V) = \frac{1}{1 + \exp\left(-\frac{V + 30}{10}\right)};$$

$$\alpha_d(V) = \frac{1.045 \times 10^{-2}(V + 35)}{1 - \exp\left(-\frac{V + 35}{2.5}\right)} + \frac{3.125 \times 10^{-2} V}{1 - \exp\left(-\frac{V}{4.8}\right)};$$

$$\beta_d(V) = \frac{-4.21 \times 10^{-3}(V - 5)}{1 - \exp\left(\frac{V - 5}{2.5}\right)};$$

$$\alpha_f(V) = \frac{-3.55 \times 10^{-4}(V + 20)}{1 - \exp\left(\frac{V + 20}{5.633}\right)}, \quad \beta_f(V) = \frac{9.44 \times 10^{-4}(V + 60)}{1 + \exp\left(-\frac{V + 29.5}{4.16}\right)};$$

$$\alpha_q(V) = \frac{3.4 \times 10^{-4}(V + 100)}{\exp\left(\frac{V + 100}{4.4}\right) - 1} + 4.95 \times 10^{-5};$$

$$\beta_q(V) = \frac{5 \times 10^{-4}(V + 40)}{1 - \exp\left(-\frac{V + 40}{6}\right)} + 8.45 \times 10^{-5};$$

$$\alpha_p(V) = \frac{9 \times 10^{-3}}{1 + \exp\left(-\frac{V + 3.8}{9.71}\right)} + 6 \times 10^{-4}, \quad \beta_p(V) = \frac{-2.25 \times 10^{-4}(V + 40)}{1 - \exp\left(\frac{V + 40}{13.3}\right)}.$$

Let us examine the dynamics and bifurcation structure by XPPAUT. The ode file of the YNI model (5.1) is shown in Fig. 5.4. When we start XPPAUT and choose the YNI.ode file, the main window will appear. If we click on the Initialconds button and then click on the (G)o button, a waveform of "V vs T" will be drawn as Fig. 5.5. Notice that, since the range where the variable T should be changed, is set to a large value of 2,000 as a default value (the statement "total=2000" in the ode file), a message "storage full" may appear. If we click on Yes (possibly several times), then we can go on the computation (or if we increase the value of dt, the messages may disappear or the number of messages may decrease). As shown in Fig. 5.5, the YNI model shows repetitive spiking (periodic solution) and its period is about 380 ms. In the following, let us examine how such a *pacemaker activity* of a heart pacemaker cell appears by using bifurcation analysis. Pacemaker activity controls the rhythmic motion of a heart. Thus, the period of the pacemaker activity is essential for the normal function of the heart and for our life. In the following, we also examine how the period varies under the changes of parameter values.

```
# YNI model YNI.ode
#
dV/dt=-1.0*(INa+IK+Il+Is+Ih-Iext)/C
dm/dt=alpha_m*(1.0-m)-beta_m*m
dh/dt=alpha_h*(1.0-h)-beta_h*h
dp/dt=alpha_p*(1.0-p)-beta_p*p
dd/dt=alpha_d*(1.0-d)-beta_d*d
df/dt=alpha_f*(1.0-f)-beta_f*f
dq/dt=alpha_q*(1.0-q)-beta_q*q
#
INa=cNa*GNa*m*m*m*h*(V-30)
IK=cK*GK*p*(exp(0.0277*(V+90))-1)/(exp(0.0277*(V+40)))
Il=cl*Gl*(1-exp((V+60)/(-20)))
Is=cs*Gs*(0.95*d+0.05)*(0.95*f+0.05)*(exp((V-10)/15)-1)
Ih=ch*Gh*q*(V+45)
#
alpha_m=(V+37)/(1-exp((V+37)/(-10)))
beta_m=40*exp((V+62)/(-17.8))
alpha_h=1.209*0.001*exp((V+20)/(-6.534))
beta_h=1/(1+exp((V+30)/(-10)))
alpha_p=9*0.001/(1+exp((V+3.8)/(-9.71)))+6*0.0001
beta_p=(-2.25*0.0001*(V+40))/(1-exp((V+40)/13.3))
alpha_d=1.045*0.01*(V+35)/(1-exp((V+35)/(-2.5)))
                          +3.125*0.01*V/(1-exp(V/(-4.8)))
beta_d=(-4.21*0.001*(V-5))/(1-exp((V-5)/2.5))
alpha_f=-3.55*0.0001*(V+20)/(1-exp((V+20)/5.633))
beta_f=(9.44*0.0001*(V+60))/(1+exp((V+29.5)/(-4.16)))
alpha_q=3.4*0.0001*(V+100)/(-1+exp((V+100)/4.4))+4.95*0.00001
beta_q=5*0.0001*(V+40)/(1-exp((V+40)/(-6)))+8.45*0.00001
#
param cNa=1, cK=1.0, cl=1.0, cs=1.0, ch=1.0
param C=1, Iext=0
param GNa=0.5, GK=0.7, Gl=0.8, Gs=12.5, Gh=0.4
V(0)=-50.0, m(0)=0.0, h(0)= 0.0
p(0)=0.0, d(0)=0.0, f(0)=0.0, q(0)= 0.0
#
@ total=2000,bound=2000,dt=0.05,dtmin=1e-6,xhi=2000
@ yhi=50,ylo=-100,meth=runge-kutta
@ autoxmin=-2,autoxmax=5,autoymin=-100,autoymax=50
@ ntst=50 nmax=250 npr=200 parmin=-2 parmax=5 dsmax=0.5 ds=0.1
#
done
```

Fig. 5.4 The YNI.ode file

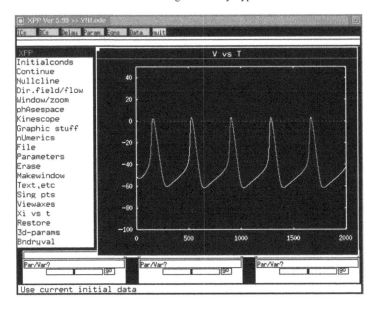

Fig. 5.5 Repetitive action potentials of the YNI model

5.2.2 Bifurcation Analysis of the YNI Model

If we want to make a one-parameter bifurcation diagram, we need to change the parameter values, since at the present parameter values, the YNI model does not possess any stable equilibrium points but have a stable periodic solution. Let us click on the Param button in the top row, then a new window will appear as Fig. 5.6. We change the value of cNa to −2.0. Click on the Initialconds button and then click on the (G)o button. After it finished, we click on the Initialconds button and then click on the (L)ast button several times in order to obtain a sufficient convergence to a stable equilibrium point. Finally, we click on the File button and then click on the Auto button, a AUTO window will appear.

If we click on Run and Steady state buttons in the AUTO window, a one-parameter bifurcation diagram will be drawn as Fig. 5.7 where the bifurcation parameter on the horizontal axis is c_{Na}. Next, click on Grab button and a cross (cursor) will appear on the one-parameter bifurcation diagram. We use Tab and then enter (return) keys to move to and grab the Hopf bifurcation (HB) point. The information about the point will be displayed below the one-parameter bifurcation diagram. Here we select the HB point whose label is 5.

Next, let us explain how to label specific solutions (for example, periodic solutions which have a specific period) in a bifurcation diagram. We click on Usr period button and select 4 button to input four user-specified conditions. A new window of AutoPer will appear and we set the values of Uzr1, ..., Uzr4, as cNa = 1.0, T = 300, T = 500, T = 700, as shown in Fig. 5.8. Finally, if we click

Fig. 5.6 The Parameters
window

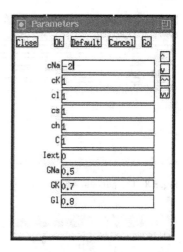

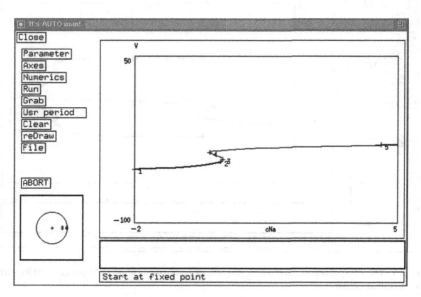

Fig. 5.7 The one-parameter bifurcation diagram of equilibrium points

on Run and Periodic buttons, a bifurcation diagram of both equilibrium points
and periodic orbits with several labels will appear as Fig. 5.9. The abscissa is the
bifurcation parameter c_{Na} and the ordinate is the membrane potential V. The points
whose labels are 10, 11, 12 and 13 are the points we have specified; at the point 11,
the values of c_{Na} is 1.0, and the periods of periodic orbits at the points 10, 12 and
13, are 300, 500 and 700, respectively. Also, we can see several information about
the solutions (periods and values of parameters) below the bifurcation diagram if
we move to the points by Grab button and Tab key.

Uzr1:cNa=1.0   Uzr6:T=26
Uzr2:T=300   Uzr7:T=29
Uzr3:T=500   Uzr8:T=32
Uzr4:T=700   Uzr9:T=35
Uzr5:T=23

Ok Cancel

Fig. 5.8 The AutoPer window

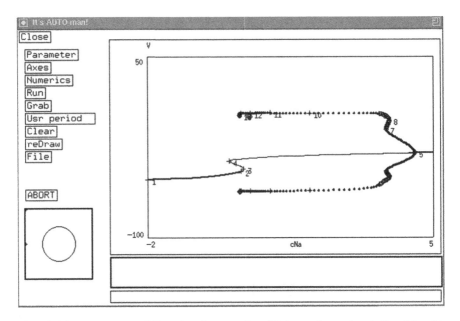

Fig. 5.9 The one-parameter bifurcation diagram of equilibrium points and periodic orbits with user-specified labels

In the bifurcation diagram of Fig. 5.9, there are two Hopf bifurcation points (labeled 2 and 5). There are also two saddle-node bifurcation points labeled 3 and 4 which are denoted as LP (Limit Point) in the AUTO. The points with labels 7 and 8 are the double-cycle bifurcation points (or saddle-node bifurcation points of periodic orbits) which are also denoted as LP in the AUTO. The pacemaker activity (repetitive spiking) bifurcates from an equilibrium points through the *right* Hopf bifurcation labeled 5. The periods of these periodic solutions increase as the parameter value of c_{Na} decreases. The pacemaker activity disappears near the left Hopf bifurcation point and the saddle-node bifurcation point 3. Therefore, pacemaker cells cannot show regular pacemaker activity if c_{Na} takes a large value or a negative value (note that the negative value is physiologically impossible because this denotes a conductance). Next, we examine how the Hopf bifurcation points move if we change the values of other parameters; we make two-parameter bifurcation diagrams.

5.2.3 Two-Parameter Bifurcation Analysis of the YNI Model

In order to make a two-parameter bifurcation diagram of the HB points, we Grab the HB point labeled as 5. Then we click on Axes and Two par buttons. A new window as shown in Fig. 5.10 will appear, and we set as follows: Ymin=-0.5, Ymax=2.5. If we click on Run and Two Param buttons, we can obtain a *part* of a two-parameter bifurcation diagram as shown in Fig. 5.11. Since the Hopf bifurcation curve immediately escapes the area of the diagram, only a small portion of bifurcation curve appears. Note that the second parameter on the vertical axis is c_K (which is preset as the default second parameter in the YNI.ode file). If we want to change this parameter, click on the Parameter button and then change the value of *Par2 from cK to any other parameters we like. In order to obtain a whole Hopf bifurcation curve, we *reverse* the tracing direction of the Hopf bifurcation curve. To do so, we click on Numerics button to change Ds to -0.1. Then, we Grab the HB point whose label is 5 again. Clicking on Run and Two Param buttons, a full part of the two-parameter bifurcation diagram is obtained as Fig. 5.12.

Next, let us calculate a two-parameter bifurcation diagram of another HB point labeled as 2 in Fig. 5.7 or Fig. 5.9. To do so, click on Grab button, then following the information below the two-parameter bifurcation diagram, we grab the HB point 2. Next, we set back the value of Ds to 0.1 using Numerics button and then click on Run and Two Param buttons. Again, change the value of Ds to -0.1, then clicking on Run and Two Param buttons will give us the two parameter bifurcation diagram

Fig. 5.10 AutoPlot window

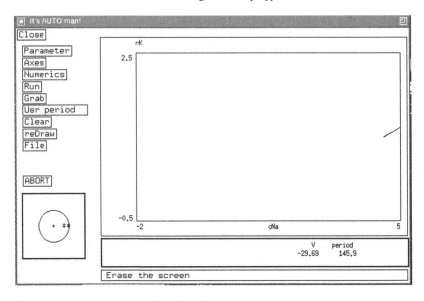

Fig. 5.11 *Partial* two-parameter bifurcation diagram

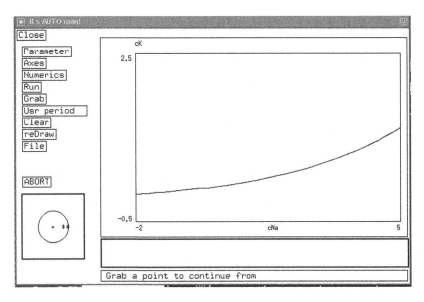

Fig. 5.12 *Whole* two-parameter bifurcation diagram

shown in Fig. 5.13. Note that in this bifurcation diagram the previous bifurcation curve traced from the HB point 5 is also superimposed (remained). In the parameter region between the two Hopf bifurcation curves, the YNI model shows repetitive spiking or pacemaker activity. Next, we examine how the period of pacemaker activity changes in such a parameter region.

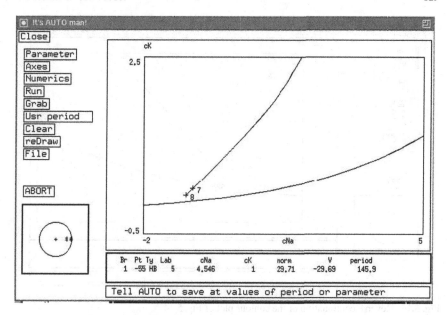

Fig. 5.13 Two Hopf bifurcation curves started from the HB points 5 and 2

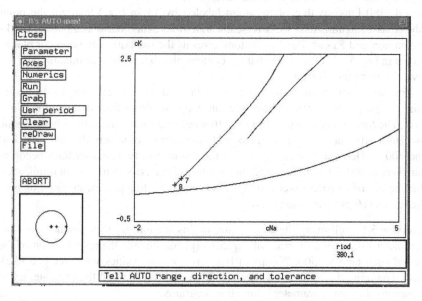

Fig. 5.14 A contour plot on periodic solutions which have a specific period of 380.1 ms

Let us draw a contour plot on the period of periodic solutions. Grab the point whose label is 11 (the point of cNa $= 1.0$) following the information below the bifurcation diagram, as explained before. Then, clicking on Run and Fixed Period buttons, a part of the contour plot will be drawn as Fig. 5.14. Along this curve

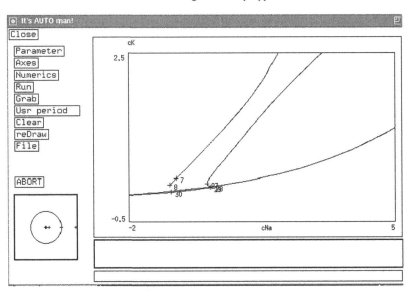

Fig. 5.15 Reverse continuation of the contour plot of Fig. 5.14

between the two Hopf bifurcation curves, all periodic solutions have the same period (380.1 ms) as that of the point labeled by 11 in Fig. 5.9. Grab the point labeled as 11 again. Also, we reverse the sign of Ds using Numerics button. Clicking on Run and Fixed Period buttons gives us the full part of the contour plot as shown in Fig. 5.15. We can see that the contour plot extends and terminates near the lower Hopf bifurcation curve.

For the other points which we have labeled as 10, 12 and 13 by using Usr period button in the above section, we can obtain the contour plots similarly. The final result is shown in Fig. 5.16 where the four contour lines between the two Hopf bifurcation curves correspond to the four specific periods: 300, 380, 500 and 700 ms (leftward). We can see that the distance of two contour lines becomes narrower near the upper Hopf bifurcation curve. This leads an important implication that the period of pacemaker activity is very sensitive to a parameter variation near the (upper) Hopf bifurcation curve.

Exercise 5.1. Following the explanation of this section, try to reproduce the two-parameter bifurcation diagram of Fig. 5.16 (please change the specification of the period such as 300, 500, 700 ms in Fig. 5.16 to other values whatever you like). Also, changing the bifurcation parameters on the abscissa and the ordinate, try to compute other two-parameter bifurcation diagrams.   □

5.2.4 Bifurcation and Parameter Sensitivity

Figure 5.17 is the more detailed bifurcation diagram of Fig. 5.16 where the bifurcation parameters are c_{Na} and c_K same as Fig. 5.16. This bifurcation diagram

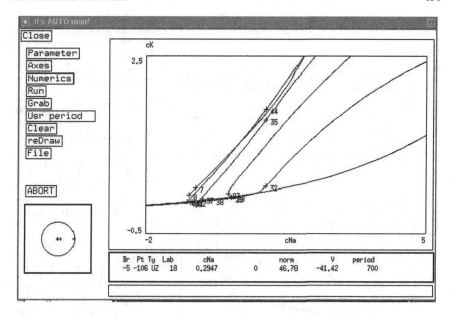

Fig. 5.16 Contour plot on periodic solutions which have specific periods: 300, 380, 500 and 700 ms

was computed by AUTO instead of XPPAUT, and more bifurcation curves are included than Fig. 5.16. The labels HB, SN, DC, PD and HC mean the bifurcation points of Hopf, saddle-node, double-cycle, period-doubling and homoclinic bifurcations, respectively. The curve labeled with "normal" denotes the contour curve of period 380 ms and other numbers denote the periods of other periodic orbits. The special point labeled as BT denotes the Bogdanov–Takens bifurcation point where three bifurcation curves of Hopf, saddle-node and homoclinic bifurcation meet together.

The bifurcation curves of HB1 and HB2 separate Fig. 5.17 into three areas. In area 2, various periodic solutions exist. When (c_{Na}, c_K) takes the values near $(-1.0, 0.0)$, periodic solutions with long period exist, and these solutions correspond to sinus bradycardia. The period becomes small when c_{Na} is increased, and it becomes large when c_K is increased. Panels (c) and (e) show two abnormal waveforms of membrane potential when c_{Na} takes a small and a large value (c_K is fixed to 1.0), respectively. If we want to get the normal period 380 ms in such abnormal cases of c_{Na}, the value of c_K should be adjusted as shown in the panels (d) and (f). Panels (a) and (b) show the typical waveforms in area 1 and area 3, respectively. Both of the membrane potentials converge to an equilibrium point eventually and cannot show repetitive action potential (thus, abnormal case).

If c_K is fixed to 1.0 and c_{Na} is varied (this corresponds to a horizontal line with $c_K = 1$ in Fig. 5.17), the "one-parameter" bifurcation diagram on the parameter c_{Na} can be obtained as shown in Fig. 5.18, in which the value of V in the steady state was plotted for each value of c_{Na}. The solid and broken curves show stable and unstable

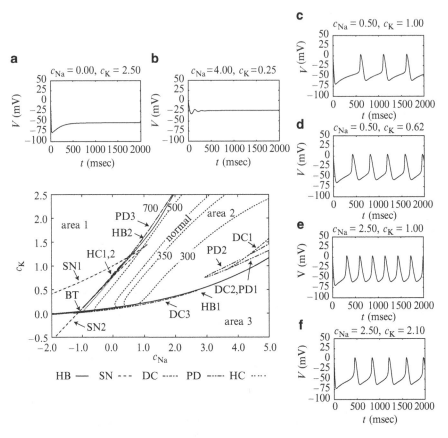

Fig. 5.17 Two-parameter bifurcation diagram of the YNI model as for the two bifurcation parameters c_{Na} and c_K. (**a–f**) Examples of membrane potential waveforms

equilibrium points, respectively. The symbols • and ○ show the maximum values of V of stable and unstable periodic solutions, respectively. Periods of periodic solutions are also shown in the diagram. In normal condition ($c_{Na} = 1.0$), a stable periodic solution whose period is about 380 ms exists, and panel (c) shows the corresponding waveform of membrane potential V.

For each value of c_{Na} between HB1 and HB2, a periodic solution (stable or unstable) exists. The period of periodic solution varies with c_{Na}. When c_{Na} is increased, the period decreases, and thus the heart rate increases. In general, a very high heart rate (>325 beats/min) corresponds to sinus tachycardia, and a very low heart rate (<130 beats/min) corresponds to sinus bradycardia (note that the YNI model is the cardiac pacemaker cell model of a rabbit). Figure 5.18b,d show two typical waveforms of membrane potential, whose periods are large and small, respectively. For the treatment of arrhythmia, it is important to consider the drug sensitivity of ion channels. In Fig. 5.18, the variation of period is small when c_{Na} is increased from

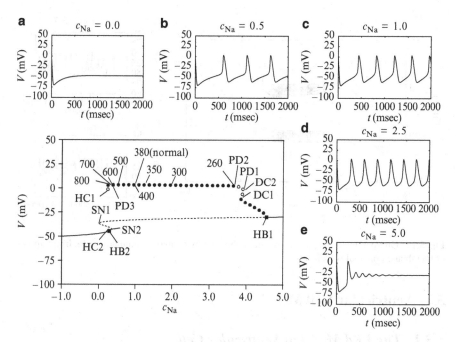

Fig. 5.18 One-parameter bifurcation diagram as for the bifurcation parameter c_{Na} obtained by AUTO. (**a–e**) Examples of membrane potential waveforms

1.0 (normal value), and it is large when c_{Na} is decreased from 1.0. In particular, when c_{Na} takes a value near 0.25, the period changes drastically. These results show that the drug sensitivity in the case of a small value of c_{Na} is stronger than that in the case of a large value of c_{Na}.

In both the left side of HB2 and the right side of HB1, only equilibrium points exist. Because of the abnormality of Na^+ channel there (c_{Na} is too small or too large), pacemaker cells cannot generate action potentials periodically and continuously. The typical waveforms of membrane potential in the two cases are shown in panels (a) and (e), respectively. Both of the membrane potentials converge to the equilibrium points, but the values of membrane potential are different in the two cases (one is low [repolarized] and the other is high [depolarized]).

Since only unstable periodic solutions and unstable equilibrium points were detected by AUTO for the values of c_{Na} between PD2 and DC1 in Fig. 5.18, we also computed the one-parameter bifurcation diagram by numerical simulations (Fig. 5.19) for the parameter values of c_{Na} between 3.6 and 4.0. In this diagram, both the local maximum and minimum values of V for each value of c_{Na} were plotted. There are many bifurcations and possibly chaotic solutions. The waveforms of membrane potentials when $c_{Na} = 3.7$ and 3.8 are shown in panels (a) and (b), respectively. In both cases, the amplitude of membrane potential varies with time, which shows serious abnormalities in the action potential generation.

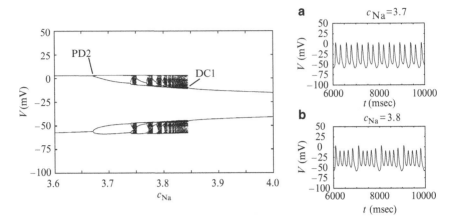

Fig. 5.19 One-parameter bifurcation diagram obtained by numerical simulations. (**a, b**) Examples of membrane potential waveforms

5.3 Ventricular Cell Model

5.3.1 The LRd Model of Ventricular Cell

The YNI model is a relatively simple and old HH-type model of cardiac pacemaker cells. In this section, we briefly explain the bifurcation structure of more *detailed* cardiac cell model. The Luo–Rudy dynamic (LRd) model (Luo and Rudy 1994) is one of such detailed cardiac cell models and is a model of cardiac ventricular cell. "Ingredients" of the LRd model are schematically illustrated in Fig. 5.20 and Table 5.1. Comparing with the YNI model of a pacemaker cell in Fig. 5.3, we can see that many factors are taken into account in the LRd model. For example, there are different types of Ca^{2+} channels such as $I_{Ca(L)}$, $I_{Ca(T)}$ and $I_{Ca,b}$, which flow the same ion Ca^{2+} but have different characteristics. Other than ionic channels, the LRd model includes "pump" and "exchanger." I_{NaCa} denotes the electric current of the Na–Ca exchanger which simultaneously moves $3Na^+$ ions *into* the cell (*down* its electrochemical gradient) and $1Ca^{2+}$ ions *out of* the cell (*up* its electrochemical gradient); as a whole, I_{NaCa} is an inward current in its normal mode of operation. Energy of the movement of Ca^{2+} against its electrochemical gradient is provided by the movement of Na^+ ions down its electrochemical gradient. I_{NaK} denotes the Na–K pump which moves Na^+ out of the cell and K^+ into the cell. Both ions move up their electrochemical gradients and energy for this movement comes from the hydrolysis of ATP.

The flow of K^+ out of the cell and the flow of Na^+ and Ca^{2+} into the cell down their electrochemical gradients through ion channels *gradually* change the intracellular ionic concentrations and destroy their concentration gradients across the cell membrane. Both Na–Ca exchanger and Na–K pump maintain the concentration gradients. (The changes of intracellular ionic concentrations are very slow. Therefore,

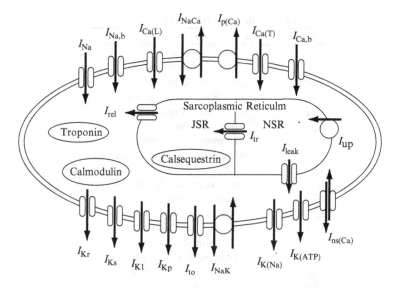

Fig. 5.20 Schematic diagram of the LRd model. This is a modification from Luo and Rudy (1994), and details are provided in the reference. This model includes many ionic channels, ionic pumps, exchangers, and Ca^{2+} buffers

in the YNI model, Na–Ca exchanger and Na–K pump are not taken into account, and the ionic concentrations are treated as constants.) In the LRd model, all ionic concentrations are considered not as constants but *variables*. In particular, the intracellular concentration of Ca^{2+} is modulated not only by ionic channels and Na–Ca exchangers but also by the "Ca^{2+} buffers" such as troponin, calmodulin and calsequestrin. Taking all ingredients in Fig. 5.20 into account, the LRd model is described by the nonlinear ordinary differential equations with 21 variables. In the same way as other HH-type models, the temporal variation of the membrane potential V(mV) is described by

$$\frac{dV}{dt} = -\frac{1}{C}(I_{\text{total}} - I_{\text{ext}}), \tag{5.2}$$

where C (μF cm^{-2}) is the membrane capacitance, I_{total} (μA cm^{-2}) the sum of all ionic currents across the membrane, and I_{ext} (μA cm^{-2}) the external stimulus current. The term I_{total} includes many ionic currents via not only ion channels but also ion exchangers and pumps, and thus the LRd model becomes 21-dimensional. Since the equations of LRd model is very complicated and lengthy, they are omitted in this book and all details of the equations can be found in Luo and Rudy (1994).

Figure 5.21 shows an action potential waveform of the LRd model, when a periodic pulse is applied. (Note that, differently from the cardiac pacemaker cell, the ventricular cell does not produce any action potentials without external stimulus.) Since it is rather difficult to analyze the bifurcation structure of the LRd model under periodic pulsatile stimulation, we use the constant direct current as an external stimulus I_{ext} in the following.

Table 5.1 Ingredients of the LRd model

Notation	Explanation
I_{Na}	Fast Na^+ current
$I_{Na,b}$	Background Na^+ current
$I_{Ca(L)}$	L-type Ca^{2+} current
I_{NaCa}	Na^+–Ca^{2+} exchange current
$I_{p(Ca)}$	Ca^{2+} pump current
$I_{Ca(T)}$	T-type Ca^{2+} current
$I_{Ca,b}$	Background Ca^{2+} current
I_{Kr}	Rapid delayed rectifier K^+ current
I_{Ks}	Slow delayed rectifier K^+ current
I_{K1}	Time-independent K^+ current
I_{Kp}	Plateau K^+ current
I_{to}	Transient outward current
I_{NaK}	Na^+–K^+ pump current
$I_{K(Na)}$	Na^+ activated K^+ current
$I_{K(ATP)}$	ATP activated K^+ current
$I_{ns(Ca)}$	Non-specific Ca^{2+} activated current
NSR	Network sarcoplasmic reticulum
JSR	Junctional sarcoplasmic reticulum
I_{up}	Ca^{2+} uptake from myoplasm to NSR
I_{tr}	Ca^{2+} transfer from NSR to JSR
I_{leak}	Ca^{2+} leak from NSR to myoplasm
I_{rel}	Ca^{2+} release from JSR to myoplasm
Troponin	
Calmodulin	Calcium buffers
Calsequestrin	

Fig. 5.21 The action potential waveform simulated by the LRd model and the periodic stimulus current I_{ext} (*upper panel*). The stimulus current is applied to the cell every 1,000 ms

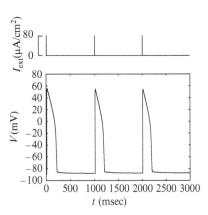

5.3.2 Bifurcation Structure of the LRd Model

Figure 5.22 is a one-parameter bifurcation diagram when $\overline{G}_{Na}$ is 600 and the other parameters are set in their normal values. The bifurcation parameter in the abscissa

Fig. 5.22 One-parameter bifurcation diagram as for the bifurcation parameter I_{ext} ($\overline{G}_{Na} = 600$)

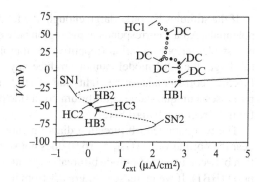

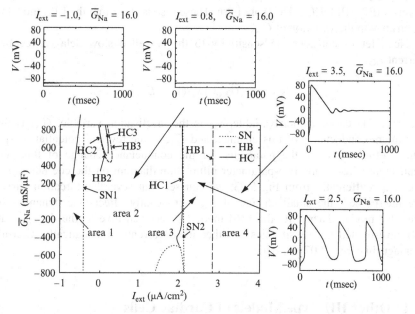

Fig. 5.23 Two-parameter bifurcation diagram on the two parameters I_{ext} and $\overline{G}_{Na}$. This figure also shows examples of a typical waveform in each area which is separated by the bifurcation curves

is the constant stimulus current I_{ext}. There are two saddle-node bifurcation points (SN1, SN2), three Hopf-bifurcation points (HB1–HB3), several double-cycle bifurcation points (DC), and three homoclinic bifurcation points (HC1–HC3).

Figure 5.23 is the two-parameter bifurcation diagram where the abscissa denotes the constant stimulus current I_{ext} and the ordinate is the maximum conductance $\overline{G}_{Na}$ of the I_{Na} current. In areas 1 and 2 where the stimulus current I_{ext} is small, the LRd model does not generate any action potentials, since ventricular cells cannot generate any action potentials without sufficiently large stimulus current. Area 1 in Fig. 5.23 corresponds to the left side of SN1 in Fig. 5.22 and area 2 to the region between SN1 and SN2 in Fig. 5.22. In both regions, the LRd model possesses a unique stable equilibrium point.

If the stimulus becomes large enough, the LRd model generates a train of action potentials. Area 3 corresponds to the region between SN2 and HB1 in Fig. 5.22, and one stable periodic solution (repetitive excitation) exists. If the stimulus becomes too large, the LRd model cannot produce any action potentials again. Area 4 in Fig. 5.23 corresponds to the right side of HB1 in Fig. 5.22 where the LRd model possesses a unique stable equilibrium point (the potential is depolarized, differently from the region leftward SN2).

The two-parameter bifurcation diagram shows that a *moderate* change of $\overline{G}_{Na}$ near the normal value ($\overline{G}_{Na} \approx 16.0$) does not significantly change the relative position between two saddle-node bifurcation points (SN1, SN2) and a Hopf bifurcation point (HB1). If we increase $\overline{G}_{Na}$ *greatly* from its normal value, four new bifurcation points (HB2, HB3, HC2, HC3) are born. These results show that the LRd model is insensitive to the variation of $\overline{G}_{Na}$.

Next, let us consider the sensitivity to the so-called slow delayed rectifier current I_{Ks}:

$$I_{Ks} = c_{Ks}\overline{G}_{Ks}x_{s1}x_{s2}(V - E_{Ks}),$$

where x_{s1} is the activation variable, x_{s2} is the inactivation variable, $\overline{G}_{Ks}$ is the maximum conductance of this channel, and E_{Ks} is the reversal potential of I_{Ks}. The coefficient c_{Ks} is the parameter for the conductance change of this current. Figure 5.24 is the two-parameter bifurcation diagram on the coefficient c_{Ks} and I_{ext}. Differently from Fig. 5.23, we can see that several bifurcation curves are intertwined; the small change of c_{Ks} significantly affects the dynamics of the LRd model. Therefore, the LRd model is very sensitive to the I_{Ks} current. We investigated the sensitivities to other ionic currents systematically. Consult Yamaguchi et al. (2007) for detail.

5.4 Other HH-Type Models of Cardiac Cells

In this chapter, we have presented only the two examples of cardiac cell models (YNI and LRd models) and analyzed their dynamics and bifurcation structures. There are, however, diverse HH-type models on the electrophysiology of various cardiac cells. HH-type models and the computational physiology on cardiac cells has started with the Noble model (Noble 1962), which is a slight modification of the original HH equations of a squid and is a model of excitable membrane of Purkinje fibres. In the late 1970s, McAllister et al. formulated a model of action potentials in Purkinje fibres (McAllister et al. 1975), and Beeler and Reuter (1977) formulated a model of mammalian ventricular myocardial cells (Beeler–Reuter model). In 1980s, the development of single channel recording technology allowed quantitative measurements of various ionic channels. DiFrancesco and Noble (1985) developed the model of Purkinje fibres. Luo and Rudy (1991) released a mammalian ventricular myocardial cell model (LRI model: Luo–Rudy I or Luo–Rudy

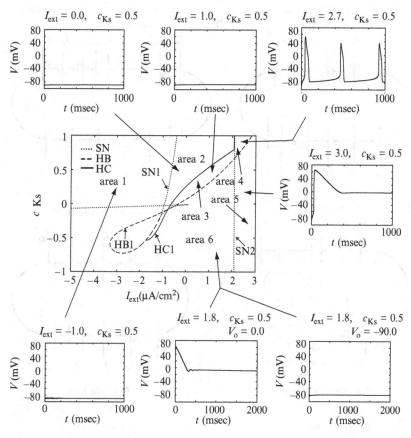

Fig. 5.24 Two-parameter bifurcation diagram on the two parameters I_{ext} and $\overline{G}_{\text{Ks}}$. This figure also shows examples of a typical waveform in each area which is separated by the bifurcation curves

phase I model) that modified the Beeler–Reuter model. The LRI model contains six ionic currents and eight variables. They also proposed the Luo–Rudy dynamic (LRd) model (Luo and Rudy 1994). The LRd model is based on recent experimental data and takes into account not only ionic channels but also ionic pumps and exchangers, and contains 21 variables as stated in the above section. Ten Tusscher et al. (2004) made the first version of human ventricular cell model, and they improved their model with more realistic description of intracellular calcium dynamics (ten Tusscher and Panfilov 2006a). Furthermore, they presented the reduced version of their human ventricular cell model (ten Tusscher and Panfilov 2006b). The list of these models which we have mentioned here is the very partial and incomplete one. In fact, there are a tremendous number of HH-type models of *cardiac cells*. Also, there is a huge repertoire of HH-type models of *other excitable cells*. For example, consult the cellML web page: http://models.cellml.org. Figure 5.25 shows a few examples of HH-type cardiac cell models other than the YNI and LRd models.

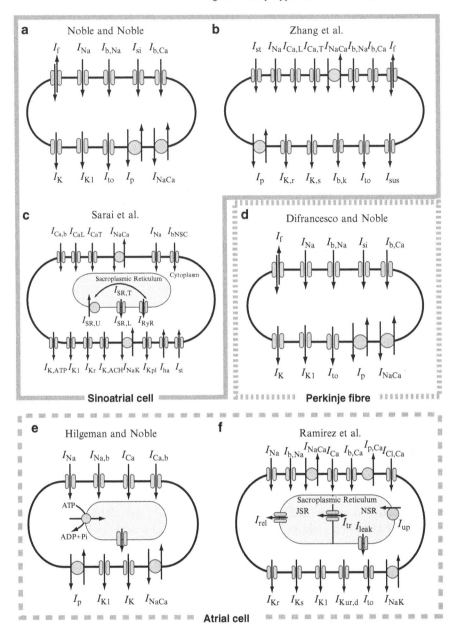

Fig. 5.25 Schematic illustrations of some HH-type cardiac cell models: (**a**) Noble and Noble (1984), (**b**) Zhang et al. (2000), (**c**) Sarai et al. (2003), (**d**) DiFrancesco and Noble (1985), (**e**) Hilgemann and Noble (1987), (**f**) Ramirez and Nattel (2000).

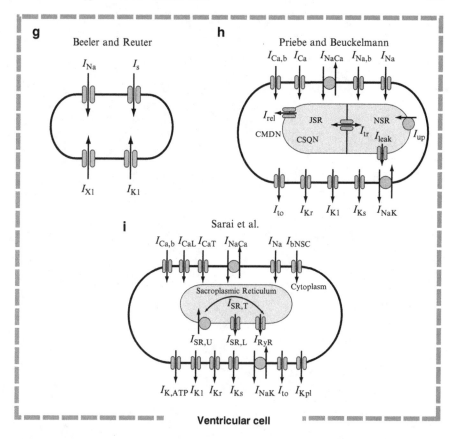

Fig. 5.25 (continued) (**g**) Beeler and Reuter (1977), (**h**) Priebe and Beuckelmann (1998), (**i**) Sarai et al. (2003). Some of these illustrations are modified with consulting the *cellML* web page: http://models.cellml.org

References

Abbott LF, Kepler TB (1990) Model neurons: from Hodgkin–Huxley to Hopfield. In: Garrido L (ed) Statistical mechanics of neural networks. Springer, Berlin

Adams P (1982) Voltage-dependent conductances of vertebrate neurones. Trends Neurosci 5:116–119

Aihara K, Matsumoto G, Ikegaya Y (1984) Periodic and non-periodic responses of a periodically forced Hodgkin–Huxley oscillator. J Theor Biol 109:249–269

Aihara K, Takabe T, Toyoda M (1990) Chaotic neural networks. Phys Lett A 144:333–340

Alexander JC, Cai DY (1991) On the dynamics of bursting systems. J Math Biol 29:405–423

Alexander JC, Doedel EJ, Othmer JC (1990) On the resonance structure in a forced excitable system. SIAM J Appl Math 50:1373–1418

Arnold L (1995) Random dynamical systems. In: Johnson R (ed) Dynamical systems. Lecture Notes in Mathematics, vol 1609. Springer, Berlin, pp 1–43

Arnold L (1998) Random dynamical systems. Springer, Berlin

Av-Ron E (1994) The role of a transient potassium current in a bursting neuron model. J Math Biol 33:71–87

Bedrov YA, Akoev GN, Dick OE (1992) Partition of the Hodgkin–Huxley type model parameter space into the regions of qualitatively different solutions. Biol Cybern 66:413–418

Beeler GW, Reuter H (1977) Reconstruction of the action potential of ventricular myocardial fibres. J Physiol (London) 268:177–210

Bertram R (1994) Reduced-system analysis of the effects of serotonin on a molluscan burster neuron. Biol Cybern 70:359–368

Bertram R, Butte MJ, Kiemel T, Sherman A (1995) Topological and phenomenological classification of bursting oscillations. Bull Math Biol 57:413–439

Braaksma B (1993) Critical phenomena in dynamical systems of van der Pol type. Thesis, Rijksuniversiteit Utrecht, Utrecht

Braaksma B (1998) Singular Hopf bifurcation in systems with fast and slow variables. J Nonlinear Sci 8:457–490

Bulsara AR, Elston TC, Doering CR, Lowen SB, Lindenberg K (1996) Cooperative behavior in periodically driven noisy integrate-fire models of neuronal dynamics. Phys Rev E 53:3958–3969

Buonocore A, Nobile AG, Ricciardi LM (1987) A new integral equation for the evaluation of first-passage-time probability densities. Adv Appl Prob 19:784–800

Butera RJ Jr (1998) Multirhythmic bursting. Chaos 8:274–284

Caianiello ER (1961) Outline of a theory of thought-processes and thinking machines. J Theor Biol 2:204–235

Canavier CC, Clark JW, Byrne JH (1991) Simulation of the bursting activity of neuron R15 in *Aplisia*: role of ionic currents, calcium balance, and modulatory transmitters. J Neurophysiol 66:2107–2124

Canavier CC, Baxter DA, Clark JW, Byrne JH (1993) Nonlinear dynamics in a model neuron provide a novel mechanism for transient synaptic inputs to produce long-term alterations of postsynaptic activity. J Neurophysiol 69:2252–2257

Carpenter GA (1977) A geometric approach to singular perturbation problems with applications to nerve impulse equations. J Diff Eqns 23:335–367

Chay TR, Keizer J (1983) Minimal model for membrane oscillations in the pancreatic β-cell. Biophys J 42:181–190

Chay TR, Rinzel J (1985) Bursting, beating, and chaos in an excitable membrane model. Biophys J 47:357–366

Chhikara RS, Folks JL (1988) The inverse Gaussian distribution: theory, methodology, and applications. M. Dekker, New York

Clay JR (1976) A stochastic analysis of the graded excitatory response of nerve membrane. J Theor Biol 59:141–158

Clay JR (1998) Excitability of the squid giant axon revisited. J Neurophysiol 80:903–913

Connor JA, Walter D, McKown R (1977) Neural repetitive firing: modifications of the Hodgkin–Huxley axon suggested by experimental results from crustacean axons. Biophys J 18:81–102

Coombes S, Bressloff PC (eds) (2005) Bursting: the genesis of rhythm in the nervous system. World Scientific, Singapore

Crill WE, Schwindt PC (1983) Active currents in mammalian central neurons. Trends Neurosci 6:236–240

Cronin J (1987) Mathematical aspects of Hodgkin–Huxley neural theory. Cambridge University Press, Cambridge

DiFrancesco D, Noble D (1985) A model of cardiac electrical activity incorporating ionic pumps and concentration changes. Philos Trans R Soc Lond [Biol] 307:353–398

Doedel E, Wang X, Fairgrieve T (1995) AUTO94 – software for continuation and bifurcation problems in ordinary differential equations. CRPC-95-2, California Institute of Technology

Doi S (1993) On periodic orbits of trapezoid maps. Adv Appl Math 14:184–199

Doi S, Kumagai S (2005) Generation of very slow neuronal rhythms and chaos near the Hopf bifurcation in single neuron models. J Comp Neurosci 19:325–356

Doi S, Sato S (1995) The global bifurcation structure of the BVP neuronal model driven by periodic pulse trains. Math Biosci 125:229–250

Doi S, Inoue J, Kumagai S (1998) Spectral analysis of stochastic phase lockings and stochastic bifurcations in the sinusoidally-forced van der Pol Oscillator with additive noise. J Stat Phys 90:1107–1127

Doi S, Inoue J, Sato S, Smith CE (1999) Bifurcation analysis of neuronal excitability and oscillations. In: Poznanski R (ed) Modeling in the neurosciences: from ionic channels to neural networks, chap 16. Harwood, Newark, NJ, pp 443–473

Fenichel N (1979) Geometric singular perturbation theory for ordinary differential equations. J Diff Eqns 31:53–98

FitzHugh R (1960) Thresholds and plateaus in the Hodgkin–Huxley nerve equations. J Gen Physiol 43:867–896

FitzHugh R (1961) Impulses and physiological states in theoretical models of nerve membrane. Biophy J 1:445–466

Fukai H, Doi S, Nomura T, Sato S (2000a) Hopf bifurcations in multiple parameter space of the Hodgkin–Huxley equations. I. Global organization of bistable periodic solutions. Biol Cybern 82:215–222

Fukai H, Nomura T, Doi S, Sato S (2000b) Hopf bifurcations in multiple parameter space of the Hodgkin–Huxley equations. II. Singularity theoretic approach and highly degenerate bifurcations. Biol Cybern 82:223–229

Gardiner CW (1983) Handbook of stochastic methods for physics, chemistry and the natural sciences. Springer, Berlin

Gerber B, Jakobsson E (1993) Functional significance of the A-current. Biol Cybern 70:109–114

Gerstein GL, Mandelbrot B (1964) Random walk models for the spike activity of a single neuron. Biophys J 4:41–68

Glass L, Mackey MC (1979) A simple model for phase locking of biological oscillators. J Math Biol 7:339–352

Glass L, Mackey MC (1988) From clocks to chaos, the rhythms of life. Princeton University Press, Princeton

Glass L, Sun J (1994) Periodic forcing of a limit-cycle oscillator: fixed points, Arnold tongues, and the global organization of bifurcations. Phys Rev E 50:5077–5084

Golomb D, Guckenheimer J, Gueron S (1993) Reduction of a channel-based model for a stomatogastric ganglion LP neuron. Biol Cybern 69:129–137

Grasman J, Jansen MJW (1979) Mutually synchronized relaxation oscillators as prototypes of oscillating systems in biology. J Math Biol 7:171–197

Guckenheimer J (1975) Isochrons and phaseless sets. J Math Biol 1:259–273

Guckenheimer J (1986) Multiple bifurcation problems for chemical reactors. Physica D 20:1–20

Guckenheimer J (1996) Towards a global theory of singularly perturbed dynamical systems. Prog Nonlinear Diff Eqns Appl 19:213–225

Guckenheimer J, Holmes P (1983) Nonlinear oscillations, dynamical systems, and bifurcation of vector fields. Springer, Berlin

Guckenheimer J, Labouriau IS (1993) Bifurcation of the Hodgkin and Huxley equations: a new twist. Bull Math Biol 55:937–952

Guckenheimer J, Gueron S, Harris-Warrick RM (1993) Mapping the dynamics of a bursting neuron. Phil Trans R Soc Lond B 341:345–359

Guevara MR, Glass L (1982) Phase locking, period-doubling bifurcations and chaos in a mathematical model of a periodically driven oscillator: a theory for the entrainment of biological oscillators and the generation of cardiac dysrhythmias. J Math Biol 14:1–23

Guttman R, Lewis S, Rinzel J (1980) Control of repetitive firing in squid axon membrane as model for neuroneoscillator. J Physiol 305:377–395

Hadeler KP, an der Heiden U, Schumacher K (1976) Generation of the nervous impulse and periodic oscillations. Biol Cybern 23:211–218

Hassard B (1978) Bifurcation of periodic solutions of the Hodgkin–Huxley model for the squid giant axon. J Theor Biol 71:401–420

Hassard BD, Shiau LJ (1989) Isolated periodic solutions of the Hodgkin–Huxley equations. J Theor Biol 136:267–280

Hata M (1982) Dynamics of Caianiello's equation. J Math Kyoto Univ 22(1):155–173

Hayashi H, Ishizuka S (1992) Chaotic nature of bursting discharges in the *Onchidium* pacemaker neuron. J Theor Biol 156:269–291

Hilgemann DW, Noble D (1987) Excitation-contraction coupling and extracellular calcium transients in rabbit atrium: reconstruction of the basic cellular mechanisms. Proc R Soc Lond B Biol Sci 230:163–205

Hille B (1992) Ionic channels of excitable membranes, 2nd edn. Sinauer, Sunderland, MA

Hodgkin AL, Huxley AF (1952) A quantitative description of membrane current and its applications to conduction and excitation in nerve. J Physiol 117:500–544

Hoppensteadt FC, Keener JP (1982) Phase locking of biological clocks. J Math Biol 15:339–349

Horikawa Y (1994) Period-doubling bifurcations and chaos in the decremental propagation of a spike train in excitable media. Phys Rev E 50:1708–1710

Inoue J, Doi S (2007) Sensitive dependence of the coefficient of variation of interspike intervals on the lower boundary of membrane potential for the leaky integrate-and-fire neuron model. Biosystems 87:49–57

Izhikevich EM (2006) Dynamical systems in neuroscience: the geometry of excitability and bursting. MIT Press, Cambridge, MA

Jones C (1996) Geometric singular perturbation theory. In: Johnson R (ed) Dynamical systems. Lecture Notes in Mathematics, vol. 1609. Springer, Berlin

Jones C, Kopell N (1994) Tracking invariant manifolds with differential forms. J Diff Eqns 108:64–88

Kawato M (1981) Transient and steady phase response curves of limit cycle oscillators. J Math Biol 12:13–30

Kawato M, Suzuki R (1978) Biological oscillators can be stopped. Topological study of a phase response curve. Biol Cybern 30:241–248

Keener JP, Glass L (1984) Global bifurcations of a periodically forced nonlinear oscillator. J Math Biol 21:175–190

Keener JP, Sneyd J (1998) Mathematical physiology. Springer, Berlin

Keener JP, Hoppensteadt FC, Rinzel J (1981) Integrate-and-fire models of nerve membrane response to oscillatory input. SIAM J Appl Math 41:503–517

Kepler TB, Marder E (1993) Spike initiation and propagation on axons with slow inward currents. Biol Cybern 68:209–214

Kepler TB, Marder E, Abbott LF (1990) The effect of electrical coupling on the frequency of model neuronal oscillators. Science 248:83–85

Kepler TB, Abbott LF, Marder E (1992) Reduction of conductance-based neuron models. Biol Cybern 66:381–387

Kokoz YuM, Krinskii VI (1973) Analysis of the equations of excitable membranes. II. Method of analysing the electrophysiological characteristics of the Hodgkin–Huxley membrane from the graphs of the zero-isoclines of a second order system. Biofizika 18:878–885

Konig P, Engel AK, Singer W (1996) Integrator or coincidence detector? The role of the cortical neuron revisited. Trends Neurosci 19:130–137

Koper MTM (1995) Bifurcation of mixed-mode oscillations in a three-variable autonomous Van der Pol–Duffing model with a cross-shaped phase diagram. Physica D 80:72–94

Krinskii VI, Kokoz YuM (1973) Analysis of the equations of excitable membranes. I. Reduction of the Hodgkin–Huxley equations to a second order system. Biofizika 18:506–511

Labouriau IS (1985) Degenerate Hopf bifurcation and nerve impulse. SIAM J Math Anal 16:1121–1133

Labouriau IS (1989) Degenerate Hopf bifurcation and nerve impulse. Part II. SIAM J Math Anal 20:1–12

Labouriau IS, Ruas MAS (1996) Singularities of equations of Hodgkin–Huxley type. Dyn Stab Syst 11:91–108

Lasota A, Mackey MC (1994) Chaos, fractals, and noise: stochastic aspects of dynamics. Springer, Berlin

Leonov NN (1959) Map of the line on to itself. Radiofisica 2:942–956

Llinas RR (1988) The intrinsic electrophysiological properties of mammalian neurons: insights into central nervous system function. Science 242:1654–1664

Luo CH, Rudy Y (1991) A model of the ventricular cardiac action potential. Depolarization, repolarization, and their interaction. Circ Res 68:1501–1526

Luo CH, Rudy Y (1994) A dynamic model of the ventricular cardiac ventricular action potential. I. Simulations of ionic currents and concentration changes. Circ Res 74:1071–1096

Maeda Y, Pakdaman K, Nomura T, Doi S, Sato S (1998) Reduction of a model for an *Onchidium* pacemaker neuron. Biol Cybern 78:265–276

Matsumoto G, Aihara K, Ichikawa M, Tasaki A (1984) Periodic and nonperiodic responses of membrane potentials in squid giant axons during sinusoidal current stimulation. J Theor Neurobiol 3:1–14

McAllister RE, Noble D, Tsien RW (1975) Reconstruction of the electrical activity of cardiac Purkinje fibres. J Physiol (London) 251:1–59

Mcdonald SW, Grebogi C, Ott E, et al (1985) Fractal basin boundaries. Physica D 17(2):125–153

Meunier C (1992) Two and three-dimensional reductions of the Hodgkin–Huxley system: separation of time scales and bifurcation schemes. Biol Cybern 67:461–468

Mira C (1987) Chaotic dynamics. World Scientific, Singapore

Mirollo RE, Strogatz SH (1990) Synchronization of pulse-coupled biological oscillators. SIAM J Appl Math 50:1645–1662

Nagumo J, Sato S (1972) On a response characteristic of a mathematical neuron model. Kybernetik 10:155–164

Nagumo J, Arimoto S, Yoshizawa S (1962) An active pulse transmission line stimulating nerve axon. Proc Inst Radio Eng 50:2061–2070

Nakano H, Saito T (2002) Basic dynamics from a pulse-coupled network of autonomous integrate-and-fire chaotic circuits. IEEE Trans Neural Netw 13:92–100

Noble D (1962) Modification of Hodgkin–Huxley equations applicable to purkinje fibre action and pace-maker potentials. J Physiol (London) 160:317–352

Noble D (1975) The initiation of the heartbeat. Oxford University Press, Oxford

Noble D (1995) The development of mathematical models of the heart. Chaos Solitons Fractals 5:321–333

Noble D, Noble SJ (1984) A model of sino-atrial node electrical activity based on a modification of the DiFrancesco–Noble (1984) equations. Proc R Soc Lond B Biol Sci 222:295–304

Nomura T, Sato S, Doi S, Segundo JP, Stiber MD (1994a) Global bifurcation structure of a Bonhoeffer van der Pol oscillator driven by periodic pulse trains. Comparison with data from an inhibitory synapse. Biol Cybern 72:55–67

Nomura T, Sato S, Doi S, Segundo JP, Stiber MD (1994b) A modified radial isochron clock with slow and fast dynamics as a model of pacemaker neurons. Global bifurcation structure when driven by periodic pulse trains. Biol Cybern 72:93–101

Okuda M (1981) A new method of nonlinear analysis for threshold and shaping actions in transient state. Prog Theor Phys 66:90–100

Pakdaman K (2001) Periodically forced leaky integrate-and-fire model. Phys Rev E 63:041907

Plant RE (1976) The geometry of the Hodgkin–Huxley model. Comp Prog Biomed 6:85–91

Poznanski RR (1998) Electrophysiology of a leaky cable model for coupled neurons. J Austral Math Soc B 40:59–71

Priebe L, Beuckelmann DJ (1998) Simulation study of cellular electric properties in heart gailure. Circ Res 82:1206–1223

Ramirez RJ, Nattel S (2000) Courtemanche M: Mathematical analysis of canine atrial action potentials: rate, regional factors, and electrical remodeling. Am J Physiol Heart Circ Physiol 279:H1767–H1785

Rescigno R, Stein RB, Purple RL, Poppele RE (1970) A neuronal model for the discharge patterns produced by cyclic inputs. Bull Math Biophys 32:337–353

Ricciardi LM (1977) Diffusion processes and related topics in biology. Springer, Berlin

Ricciardi LM, Sato S (1988) First-passage-time density and moments of the Ornstein–Uhlenbeck process. J Appl Prob 25:43–57

Rinzel J (1978) On repetitive activity in nerve. Fed Proc 37:2793–2802

Rinzel J (1985) Excitation dynamics: insights from simplified membrane models. Fed Proc 44:2944–2946

Rinzel J (1990) Discussion: electrical excitability of cells, theory and experiment: review of the Hodgkin–Huxley foundation and update. Bull Math Biol 52:5–23

Rinzel J, Keener JP (1983) Hopf bifurcation to repetitive activity in nerve. SIAM J Appl Math 43:907–922

Rinzel J, Miller RN (1980) Numerical calculation of stable and unstable periodic solutions to the Hodgkin–Huxley equations. Math Biosci 49:27–59

Rocşoreanu C, Georgescu A, Giurgiţeanu N (2000) The Fitzhugh–Nagumo model: bifurcation and dynamics. Springer, Berlin

Rush ME, Rinzel J (1994) Analysis of bursting in a thalamic neuron model. Biol Cybern 71:281–291

Rush ME, Rinzel J (1995) The potassium A-current, low firing rates and rebound excitation in Hodgkin–Huxley models. Bull Math Biol 57:899–929

Sarai N, Matsuoka S, Kuratomi S, Ono K, Noma A (2003) Role of individual ionic current systems in the SA node hypothesized by a model study. Jpn J Physiol 53:125–134

Scharstein H (1979) Input–output relationship of the leaky-integrator neuron model. J Math Biol 8:403–420

Schweighofer N, Doya K, Kawato M (1999) Electrophysiological properties of inferior olive neurons: a compartmental model. J Neurophysiol 82:804–817

Shadlen MN, Newsome WT (1994) Noise, neural codes and cortical organization. Curr Opin Neurobiol 4:569–579

Shadlen MN, Newsome WT (1998) The variable discharge of cortical neurons: implications for connectivity, computations, and information coding. J Neurosci 18:3870–3896

Shiau LJ, Hassard BD (1991) Degenerate Hopf bifurcation and isolated periodic solutions of the Hodgkin–Huxley model with varying sodium ion concentration. J Theor Biol 148:157–173

Softky WR, Koch C (1993) The highly irregular firing of cortical cells is inconsistent with temporal integration of random EPSPs. J Neurosci 13:334–350

Stein RB, French AS, Holden AV (1972) The frequency response, coherence, and information capacity of two neuronal models. Biophys J 12:295–322

Strassberg AF, DeFelice LJ (1993) Limitations of the Hodgkin–Huxley formalism: effects of single channel kinetics upon transmembrane voltage dynamics. Neural Comput 5:843–855

Takahashi N, Hanyu Y, Musha T, Kubo R, Matsumoto G (1990) Global bifurcation structure in periodically stimulated giant axons of squid. Physica D 43:318–334

Tateno T, Doi S, Sato S, Ricciardi LM (1995) Stochastic phase-lockings in a relaxation oscillator forced by a periodic input with additive noise: a first-passage-time approach. J Stat Phys 78:917–935

ten Tusscher KHW, Panfilov AV (2006a) Alternans and spiral breakup in a human ventricular tissue model. Am J Physiol Heart Circ Physiol 291:H1088–H1100

ten Tusscher KHW, Panfilov AV (2006b) Cell model for efficient simulation of wave propagation in human ventricular tissue under normal and pathological conditions. Phys Med Biol 51:6141–6156

ten Tusscher KHW, Noble D, Noble PJ, Panfilov AV (2004) A model for human ventricular tissue. Am J Physiol Heart Circ Physiol 286:H1573–H1589

Terman D (1991) Chaotic spikes arising from a model of bursting in excitable membranes. SIAM J Appl Math 51:1418–1450

Torikai H, Saito T (1999) Return map quantization from an integrate-and-fire model with two periodic inputs. IEICE Trans Fundam E82-A:1336–1343

Traub RD, Wong RKS, Miles R, Michelson H (1991) A model of a CA3 hippocampal pyramidal neuron incorporating voltage-clamp data on intrinsic conductances. J Neurophysiol 66:635–650

Troy WC (1978) The bifurcation of periodic solutions in the Hodgkin–Huxley equations. Q Appl Math 36:73–83

Tsumoto K, Yoshinaga T, Aihara K, Kawakami H (2003) Bifurcations in synaptically coupled Hodgkin–Huxley neurons with a periodic input. Int J Bifurcat Chaos 13:653–666

Tsumoto K, Kitajima H, Yoshinaga T, Aihara K, Kawakami H (2006) Bifurcations in Morris–Lecar neuron model. Neurocomputing 69:293–316

Tuckwell HC (1988) Introduction to theoretical neurobiology. Cambridge University Press, Cambridge

Tuckwell HC (1989) Stochastic processes in the neurosciences. SIAM, Philadelphia

van der Pol B (1926) On "relaxation-oscillations". Phil Mag 2:978–992

Winfree AT (1980) The geometry of biological time. Springer, New York

Xu J-X, Jiang J (1996) The global bifurcation characteristics of the forced van der Pol oscillator. Chaos Solitons Fractals 7:3–19

Yamaguchi R, Doi S, Kumagai S (2007) Bifurcation analysis of a detailed cardiac cell model and drug sensitivity of ionic channels. In: Proc. 15th IEEE international workshop on Nonlinear Dynamics of Electronic Systems 2007, pp 205–208

Yanagida E (1985) Stability of fast traveling pulse solutions of the FitzHugh–Nagumo equations. J Math Biol 22:81–104

Yanagida E (1987) Branching of double pulse solutions from single pulse solutions in nerve axon equations. J Diff Eqns 66:243–262

Yanagida E (1989) Stability of double-pulse solutions in nerve axon equations. SIAM J Appl Math 49:1158–1173

Yanagihara K, Noma A, Irisawa H (1980) Reconstruction of sinoatrial node pacemaker potential based on the voltage clamp experiments. Jpn J Physiol 30:841–857

Yellin E, Rabinovitch A (2003) Properties and features of asymmetric partial devil's staircases deduced from piecewise linear maps. Phys Rev E 67:016202

Yoshinaga T, Sano Y, Kawakami H (1999) A method to calculate bifurcations in synaptically coupled Hodgkin–Huxley equations. Int J Bifurcat Chaos 9:1451–1458

Yu X, Lewis ER (1989) Studies with spike initiators: linearization by noise allows continuous signal modulation in neural networks. IEEE Trans Biomed Eng 36:36–43

Zhang H, Holden AV, Kodama I, Honjo H, Lei M, Varghese T, Boyett MR (2000) Mathematical models of action potential in the periphery and center of the rabbit sinoatrial node. Am J Physiol 279:H397–H421

Index